NMR Spectroscopy and Polymer Microstructure: The Conformational Connection

Alan E. Tonelli

A T & T Bell Laboratories

o361053 6

CHEMISTRY

Alan E. Tonelli
AT&T Bell Laboratories
Polymer Chemistry Research Department
Room 1D-248
600 Mountain Avenue
Murray Hill, NJ 07974

Library of Congress Cataloging-in-Publication Data

Tonelli, Alan E.
  NMR spectroscopy and polymer microstructure : the conformational
  connection / Alan E. Tonelli.

      p.   cm.
  Includes bibliographical references.
  ISBN 0-89573-737-X
  1. Polymers—Analysis.   2. Nuclear magnetic resonance
  spectroscopy.   I. Title.
QD139.P6T66   1989
547.7'046—dc20   89-16640
                                                        CIP

British Library Cataloguing in Publication Data
Tonelli, Alan E.
  NMR spectroscopy and polymer microstructure.
  1. Polymers. Structure. Determination. Nuclear
  magnetic resonance spectroscopy
  I. Title
  547.7'0448

  ISBN 0-89573-737-X

Printed in the United States of America.
ISBN 0-89573-737-X VCH Publishers
ISBN 3-527-27942-3 VCH Verlagsgesellschaft

Printing History:
10 9 8 7 6 5 4 3 2 1

Published jointly by:

VCH Publishers, Inc.	VCH Verlagsgesellschaft mbH	VCH Publishers (UK) Ltd.
220 East 23rd Street	P.O. Box 10 11 61	8 Wellington Court
Suite 909	D-6940 Weinheim	Cambridge CB1 1HW
New York, New York 10010	Federal Republic of Germany	United Kingdom

Contents

Preface

Since the 1930s, when the macromolecular hypothesis championed by Staudinger finally achieved widespread scientific acceptance, polymer scientists have synthesized a multitude of long-chain, high-molecular-weight polymers and have isolated many naturally occurring polymers (proteins, polynucleotides, etc.) as well. The most fundamental and usually the information first sought concerning a polymer is its microstructure. Questions like "what is the primary sequence of amino acid residues in a protein?" or "are the phenyl rings in polystyrene attached in a stereoregular manner to the backbone?" must be posed and confronted before the physical properties of polymers can begin to be understood.

Aside from x-ray diffraction studies of stereoregular, crystallizable polymers, there were no methods for the direct experimental determination of polymer microstructure until the techniques of high-resolution NMR were applied. Of the two nuclei ^1H and ^{13}C, which possess spin and are common in synthetic polymers, ^1H initially served as the spin probe in NMR polymer studies. However, though ^1H is more abundant than ^{13}C, ^1H NMR spectra of synthetic polymers suffer from a narrow dispersion of chemical shifts and extensive $^1H-^1H$ spin–spin coupling. ^{13}C NMR as currently practiced does not suffer from these difficulties, the latter of which recently has been turned into an advantage for ^1H NMR by 2D techniques and is most successfully employed to study the microstructures of proteins.

The advent of proton-decoupled spectra recorded in the Fourier-transform mode has quickly made ^{13}C NMR spectroscopy the method of choice for determining the microstructures of synthetic polymers. The field strength at which a given ^{13}C nucleus resonates depends sensitively on its local molecular environment or microstructure. Through connection of the observed resonant frequencies with the various carbon nuclei in a polymer, its microstructure may be detailed. It is the establishment and utilization of the connections between the resonances observed in high-resolution ^{13}C NMR spectra of polymers and their microstructures which constitutes the primary subject of this volume.

It is our purpose to demonstrate the techniques, both experimental and theoretical, used to assign the high-resolution ^{13}C NMR spectra of polymers observed both in solution and in the solid. Brief descriptions of the pulsed techniques used to obtain high-resolution solution spectra in the Fourier-transform mode of operation are presented, including the INEPT, DEPT, and 2D-NMR COSY and NOESY techniques. The methods of cross-polarization, high-power proton decoupling, and magic-angle spinning used to achieve high-resolution solid-state spectra are also mentioned. Application of these techniques to determine the microstructures of synthetic homo- and copolymers and the structures of biological macromolecules are presented in the various chapters.

A common theme underlying the microstructural interpretation of polymer spectra is the connection between the local polymer conformations and the ^{13}C chemical shifts observed for the carbon nuclei in different microstructures. This connection is provided by the γ-gauche effect which results in the shielding of a ^{13}C nucleus by its nonprotonic γ-substituents when they are in a gauche arrangement. Knowledge of the local bond rotational conformations and their dependence on microstructure provides, via the γ-gauche effect, a means to predict the microstructural dependence of ^{13}C chemical shifts observed in polymers. This approach substantially aids the assignment of polymer ^{13}C NMR spectra to their microstructures.

In addition, the γ-gauche effect method of predicting ^{13}C NMR chemical shifts provides a method for determining the local conformational characteristics of polymers, whether the conformations are dynamically averaged in solution or are static in the solid. Several examples of solution and solid-state studies of polymer conformations are found in this volume, including the description of 2D NOESY ^{1}H NMR techniques for establishing the solution conformations of flexible and rigid polymers.

Most of the examples discussed are taken from the work of the author and his colleagues. Aside from familiarity, the principal reason for this parochial selection stems from the approach followed in these studies. Because of the conformational connection between the local microstructural environments and the ^{13}C chemical shifts exhibited by the carbon nuclei in polymers, these studies were based on a conformational point of view. Either the previously established conformational characteristics of polymers and their dependence on microstructure were used to aid the assignment of their ^{13}C NMR spectra to their underlying microstructures, or the conformations of polymers were derived from their experimentally established microstructure–^{13}C-NMR correlations.

It is hoped that polymer scientists who read this book will appreciate and apply some of the techniques described here, for they serve to make NMR spectroscopy a most valuable tool in the study of polymer microstructures and conformations, as found both in solution and in the solid states.

The Microstructure of Polymer Chains

1.1. Introduction

Chemistry is the application of the scientific method to the study of molecules. Which atoms constitute a particular molecule, how are they connected or bonded to each other, and what is the three-dimensional shape of this collection of atoms called a molecule? Answers to these questions yield information concerning the molecular composition, configuration, and conformation. It is just such information that we also seek when determining the microstructures of polymers.

Polymer microstructure is usually determined during the course of polymerization of monomers to form the long-chain molecules known as polymers. The direction of monomer addition, the stereochemical and geometrical isomeric forms of the incorporated monomers, and the order of addition of comonomer units to form a copolymer constitute the resultant polymer microstructures. We will briefly illustrate these microstructural features using vinyl and diene polymers as examples.

On a larger scale, sometimes as a direct result of polymerization and at other times due to postpolymerization chemical reactions, the linear architecture of polymers can be modified to yield branched and cross-linked structures. The types, lengths, and locations of the branches and crosslinks can also be considered components of a polymer's microstructure.

Finally we conclude with a brief discussion of the relation between the physical properties exhibited by polymers and their microstructures. Emphasis is placed on the wide range of properties observed for polymers and their underlying causes, the rich variety of polymer microstructures.

1.2. Polymers Are Macromolecules

Though taken for granted by present-day chemists, the fact that polymers are long-chain molecules or macromolecules with molecular weights sometimes exceeding 10^6 has only been appreciated for the past 60 years. The particular efforts of Staudinger (1932) and Carothers (1931) served to dispel the then prevalent notion that polymers were aggregates of small molecules held together by "secondary valence forces." The fascinating story of the development of the macromolecular concept of polymers is ably chronicled in the first chapter of Flory's classic polymer text (Flory, 1953) and by Morawetz (1985) in a recently published history of polymer science.

Once the long-chain nature of polymers had been established, study of the microstructures of naturally occurring and synthetic polymers began in earnest. After all, if polymers are macromolecules, then they are properly in the chemist's domain of scientific inquiry. Chemists have been creating and studying new polymer microstructures ever since.

1.3. Polymer Microstructures from Polymerization of Monomers

1.3.1. Directional Isomerism

Let us consider the types of polymer microstructures which can be produced by the isomerism of monomers during their incorporation into the growing polymer chain:

$$n \ \underset{B}{\overset{A}{\diagdown}} C = C \underset{H}{\overset{H}{\diagup}} \longrightarrow \left(\underset{\underset{B}{|}}{\overset{\overset{A}{|}}{C}} - \underset{\underset{H}{|}}{\overset{\overset{H}{|}}{C}} \right)_n$$

If both $A, B \neq H$ in the vinyl polymerization indicated above, then each monomer unit can be potentially enchained in either of two directions, i.e.

$$-\underset{\underset{B}{|}}{\overset{\overset{A}{|}}{C}} - CH_2 - \quad \text{or} \quad -CH_2 - \underset{\underset{B}{|}}{\overset{\overset{A}{|}}{C}} -, \text{ leading to among others the following}$$

regiosequences:

$$(1) \quad -\underset{B}{\overset{A}{|}}{C} - CH_2 - \underset{B}{\overset{A}{|}}{C} - CH_2 - \underset{B}{\overset{A}{|}}{C} - CH_2 - \underset{B}{\overset{A}{|}}{C} - CH_2 - \underset{B}{\overset{A}{|}}{C} - CH_2 -$$

$$(2) \quad -\underset{B}{\overset{A}{|}}{C} - CH_2 - \underset{B}{\overset{A}{|}}{C} - CH_2 - CH_2 - \underset{B}{\overset{A}{|}}{C} - \underset{B}{\overset{A}{|}}{C} - CH_2 - \underset{B}{\overset{A}{|}}{C} - CH_2 -$$

Note that in (1) all monomers units have been added in the same direction, producing a regioregular structure, while in (2) the third monomer has been inserted in the inverted direction, resulting in a regioirregular structure.

1.3.2. Stereochemical Isomerism

Let us assume that all monomers $\begin{matrix} A \\ \diagdown \\ B \diagup \end{matrix} C = C \begin{matrix} \diagup H \\ \diagdown H \end{matrix}$ are added in the same direction during their polymerization (regioregular addition). There still remains a degree of structural freedom which is fixed during the polymerization, i.e. the stereochemical configuration or relative handedness of successive monomer units. Even though the main-chain substituted carbons do not have the necessary four different substituents (disregarding chain ends) to qualify as asymmetric centers, they do have the opportunity for relative handedness and are therefore termed *pseudoasymmetric*. The three structures below illustrate the various possible stereochemical arrangements produced by the polymerization of monomers possessing a pseudoasymmetric carbon:

(1) ISOTACTIC

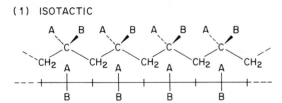

(2) SYNDIOTACTIC

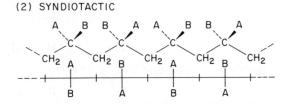

(3) ATACTIC

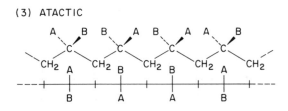

Each of the polymer chains has been drawn in the planar, zigzag, or all-*trans* conformation to provide the clearest perspective of the relative handedness of stereochemical arrangements of neighboring repeat units. The two stereoregular structures (1), (2), where the substituents A and B are all either on the same side of the zigzag backbone plane or alternate from side to side, are called *isotactic* and *syndiotactic*, respectively. The atactic structure (3) is characterized by an irregular, random arrangement of neighboring substituent groups on either side of the backbone.

Natta (1955) demonstrated, through use of the coordination catalyst developed by Ziegler for the polymerization of ethylene, that α-olefins could be polymerized to stereoregular structures, with both isotactic and syndiotactic polymers produced. This discovery was an important development in both the science and the technology of polymers, because the physical properties of stereoregular vinyl polymers are dramatically different from those exhibited by their atactic counterparts (*vide infra*).

Manipulation of molecular models will verify the fact that these stereoisomeric forms cannot be interconverted by the rotation of substituent groups around the main-chain carbon–carbon bonds. However, if either substituent A or B is a proton, it may be possible to alter the configurations of the pseudoasymmetric carbons via the reversible removal of protons, viz.

$$\sim CH_2-\underset{R}{\overset{H}{\underset{|}{\overset{|}{C}}}}\sim \underset{+H^*}{\overset{-H^*}{\rightleftharpoons}} \sim CH_2-\underset{R}{\overset{}{\underset{|}{\overset{|}{C^*}}}}\sim \rightleftharpoons \sim CH_2-\underset{*}{\overset{R}{\underset{|}{\overset{|}{C}}}}\sim \underset{-H^*}{\overset{+H^*}{\rightleftharpoons}} \sim CH_2-\underset{H}{\overset{R}{\underset{|}{\overset{|}{C}}}}\sim$$

This process of altering the stereochemical arrangements of the neighboring repeat units in a vinyl polymer is termed *epimerization* (Flory, 1967; Clark, 1968) and has been utilized to convert several isotactic vinyl polymers (Shepherd et al., 1979; Suter and Neuenschwander, 1981; Dworak et al., 1985) into their atactic counterparts. (See Chapter 6 for an example of vinyl polymer epimerization.)

If a vinyl monomer is 1,2-disubstituted, $\overset{H}{\underset{A}{\diagdown}}C=C\overset{H}{\underset{B}{\diagup}}$ or $\overset{H}{\underset{A}{\diagdown}}C=C\overset{B}{\underset{H}{\diagup}}$, each backbone carbon in the resultant polymer is pseudoasymmetric and pairwise nonidentical when A ≠ B. Below are shown three stereoregular sequences:

(1) ERYTHRODIISOTACTIC

(2) THREODIISOTACTIC

(3) DISYNDIOTACTIC

1.3.3. Geometrical Isomerism

Polymerization of diene monomers, such as butadiene (CH_2=CH—CH=CH_2), can produce polymers with varying geometrical structures. 1,4-enchainment can produce *cis*(Z) and *trans*(E) structures:

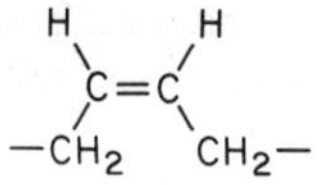

$$
\begin{array}{cc}
\underset{-CH_2}{\overset{H}{\diagdown}}\!C\!=\!C\!\underset{CH_2-}{\overset{H}{\diagup}} & \underset{-CH_2}{\overset{H}{\diagdown}}\!C\!=\!C\!\underset{H}{\overset{CH_2-}{\diagup}}
\end{array}
$$

1,4—CIS(Z) 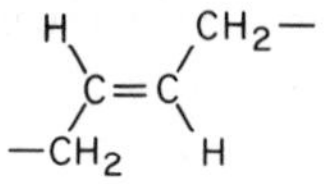   1,4—TRANS(E)

1,2-enchainment leads to structures with the same configurational properties as vinyl polymers:

(ATACTIC 1,2–POLYBUTADIENE)

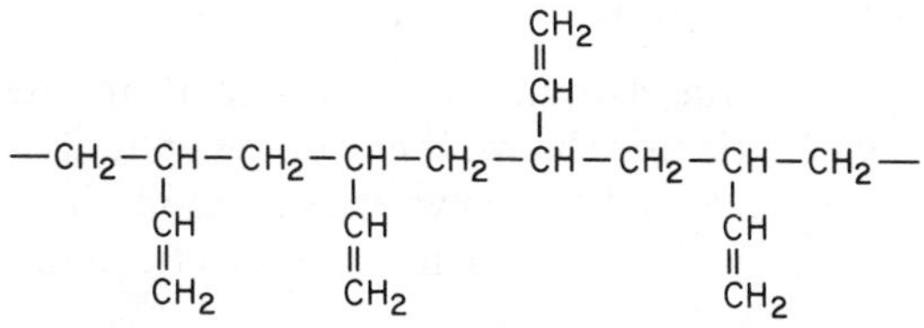

$$
\begin{array}{c}
\overset{\displaystyle CH_2}{\underset{\displaystyle CH}{\|}} \\
\end{array}
$$

$$-CH_2-CH-CH_2-CH-CH_2-CH-CH_2-CH-CH_2-$$

with pendant groups:

$$
\begin{array}{ccccc}
CH & & CH & & CH \\
\| & & \| & & \| \\
CH_2 & & CH_2 & & CH_2
\end{array}
$$

Natural rubber is an example of a polymer obtained through polymeriza-
tion of a 2-substituted butadiene, $CH_2 = \overset{\displaystyle X}{\overset{\displaystyle |}{C}} - CH = CH_2$, where $X = CH_3$
and all units correspond to a 1,4-*cis* (Z) enchainment. 1,4-*trans* (E) enchain-
ment results in balata or gutta percha, which is derived from a different plant.
1,4-enchainment of 2-substituted butadienes may result in head-to-tail
(regioregular) or head-to-head : tail-to-tail (regioirregular) structures:

Head-to-tail,

$$---CH_2 - \overset{\displaystyle X}{\overset{\displaystyle |}{C}} = CH - CH_2 - CH_2 - \overset{\displaystyle X}{\overset{\displaystyle |}{C}} = CH - CH_2 ---$$

Head-to-head : tail-to-tail,

$$---CH_2 - CH = \overset{\displaystyle X}{\overset{\displaystyle |}{C}} - CH_2 - CH_2 - \overset{\displaystyle X}{\overset{\displaystyle |}{C}} = CH - CH_2 - CH_2 - CH = \overset{\displaystyle X}{\overset{\displaystyle |}{C}} - CH_2 ---$$

The 2-substituted butadiene units may also be incorporated by 1,2 or 3,4
additions,

$$---CH_2 - \overset{\displaystyle X}{\underset{\displaystyle H_2C = CH}{\overset{\displaystyle |}{C}}} --- \quad \text{or} \quad ---CH_2 - \underset{\displaystyle H_2C = CX}{CH} ---,$$

both of which may occur in isotactic or syndiotactic stereosequences and
regular (head-to-tail) or irregular (head-to-head : tail-to-tail) regiosequences.

Dienes of the type $\overset{\displaystyle X}{\overset{\displaystyle |}{CH}} = CH - \overset{\displaystyle X}{\overset{\displaystyle |}{CH}} = CH$ lead to different stereoisomeric
structures even if they are incorporated exclusively via 1,4-addition:

$$---\overset{\displaystyle X}{\overset{\displaystyle |}{CH}} - \overset{\displaystyle X}{\overset{\displaystyle |}{CH}} - CH = CH - \overset{\displaystyle X}{\overset{\displaystyle |}{CH}} - \overset{\displaystyle X}{\overset{\displaystyle |}{CH}} - CH = CH - \overset{\displaystyle X}{\overset{\displaystyle |}{CH}} - \overset{\displaystyle X}{\overset{\displaystyle |}{CH}} ---$$

We may obtain *trans*- or *cis*-meso and *trans*- or *cis*-D,L structures. If the
1,4-substituents are different (X, Y), then even more elaborate structures may
be produced.

1.3.4. Truly Asymmetric Polymers

Vinyl monomers with true asymmetric centers in their side chains produce
polymers with asymmetric side chains. If one of the optically active monomeric
enantiomers (D or L, or R or S) is polymerized, the resulting polymer will be
optically active and may also be isotactic, syndiotactic, or atactic in the usual
sense.

Though polymerization of vinyl monomers cannot result in polymers with truly asymmetric main-chain carbons, other monomers, such as propylene oxide, do produce such polymers (Pruitt and Baggett, 1955; Price and Osgan, 1956; Osgan and Price, 1959; Tsuruta, 1967):

$$
\underset{R,\,S}{\overset{\text{CH}_3\qquad\quad H}{\underset{H\quad O\quad H}{C\!=\!=\!C}}} \longrightarrow
$$

$$
\overset{\text{CH}_3}{\underset{H}{-\,C}} - \text{CH}_2 - O - \overset{H}{\underset{\text{CH}_3}{C}} - \text{CH}_2 - O - \overset{\text{CH}_3}{\underset{H}{C}} - \text{CH}_2 - O -
$$

ISOTACTIC, RRR OR SSS

$$
\overset{\text{CH}_3}{\underset{H}{-\,C}} - \text{CH}_2 - O - \overset{\text{CH}_3}{\underset{H}{C}} - \text{CH}_2 - O - \overset{\text{CH}_3}{\underset{H}{C}} - \text{CH}_2 - O -
$$

SYNDIOTATIC, RSR OR SRS

Because the poly(propylene oxide) chains have a sense of direction, there are two heterotactic structures which cannot be superimposed (non-mirror-images):

$$
\overset{\text{CH}_3}{\underset{H}{-\,C}} - \text{CH}_2 - O - \overset{H}{\underset{\text{CH}_3}{C}} - \text{CH}_2 - O - \overset{H}{\underset{\text{CH}_3}{C}} - \text{CH}_2 - O -
$$

HETEROTACTIC-1 RRS OR SSR

$$
\overset{H}{\underset{\text{CH}_3}{-\,C}} - \text{CH}_2 - O - \overset{H}{\underset{\text{CH}_3}{C}} - \text{CH}_2 - O - \overset{\text{CH}_3}{\underset{H}{C}} - \text{CH}_2 - O -
$$

HETEROTACTIC-2 SRR OR RSS

In nature the most prevalent examples of truly asymmetric polymers are the polypeptides and proteins $\left(\!\!-\text{NH}-\overset{R}{\underset{}{\text{CH}}}-\overset{O}{\underset{}{\text{C}}}-\!\!\right)_{\!n}$. In proteins the peptide residues are invariably of the L-configuration, while in small, usually cyclic polypeptides which function as hormones, toxins, antibiotics, or ionophores (Tonelli, 1986) both L and D residues are found.

1.3.5. Copolymer Sequences

To this point we have limited our discussion of polymer microstructure to homopolymers obtained by polymerization of a single monomer. However, as is common in nature (e.g. proteins and polynucleic acids), two or more different monomer units may be incorporated, resulting in a copolymer. The

Figure 1.1 ■ Copolymer types.

comonomer units may be incorporated at random, in regular alternation, or in block or graft structures. (See Figure 1.1.)

Aside from the comonomer sequence and method of attachment, copolymer microstructure is also subject to the effects of stereo- and regiosequence as discussed previously. Clearly the microstructures of homo- and copolymers are varied and afford the synthetic chemist a limitless number of variations on a macromolecular theme.

1.4. Organization of Polymer Chains

Although our discussion may have implied otherwise, polymer chains are not necessarily linear (see Figure 1.2). Polymers may contain branches produced during their polymerization or grafted onto their backbones by postpolymerization reaction. Polymerization of multifunctional monomers (trifunctional and higher) or postpolymerization reaction and/or irradiation can produce polymer chains that are cross-linked. At sufficiently high degrees of cross-linking a three-dimensional polymeric network is formed.

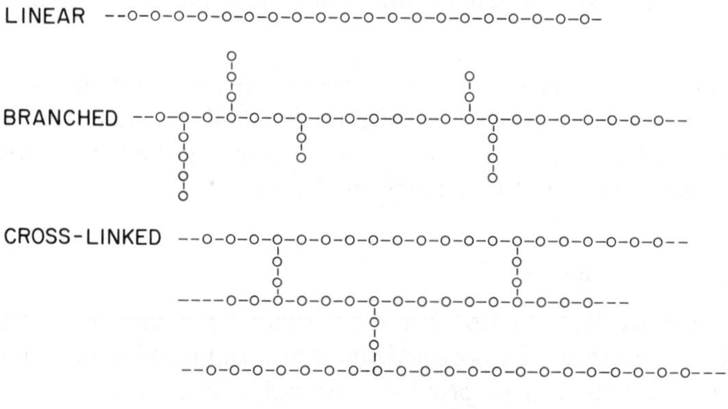

LINEAR --o-o-o-o-o-o-o-o-o-o-o-o-o-o-o-o-o-o-o-o-o-o-o-o-

BRANCHED --o-o-o-o-o-o-o-o-o-o-o-o-o-o-o-o-o-o-o-o-o-o-o-o--

CROSS-LINKED --o-o-o-o-o-o-o-o-o-o-o-o-o-o-o-o-o-o-o-o-o-o-o--

Figure 1.2 ■ Polymer chain structures.

1.5. Polymer Properties and Their Relation to Microstructure

Polymer scientists are motivated in their studies of polymer microstructure by the desire to understand the amazingly varied and often unique physical properties shown by this class of macromolecular materials. Why are some polymers rigid and strong, while others are elastic and deformable and may even flow? Why do some polymers resist degradation caused by exposure to heat, radiation, and chemicals, while others rapidly degrade? Why do some polymers become brittle and shatter on impact at low temperatures, while others are tough and retain their resistance to impact under the same conditions? What is it about the microstructures of proteins (primary sequence of peptide residues) that make some proteins structural components of teeth, bones, skin, and hair, while other proteins function as enzymes to catalyze the body's biochemical reactions? Answers to these and similar questions are most likely to result from a detailed knowledge of polymer microstructure.

Let us briefly illustrate the connections between polymer properties and polymer microstructure. It is now well known (Ward, 1985) that the strongest polymers are those whose linear chains can be highly oriented into fibers where they are crystalline. The ability of a polymer chain to crystallize depends on its microstructure. For example, polypropylene

$$CH_3$$
$$|$$
$[-(CH_2-CH-)_n]$ (PP), when polymerized with the appropriate catalyst, is highly isotactic (i). The stereoregular placement of pendant methyl groups in i-PP permits the facile crystallization of its chains ($T_{melt} = 200°C$). When crystalline i-PP chains are drawn, strong fibers are produced, which can be used for example to produce strong, lightweight rope. Atactic PP, with its stereoirregular distribution of methyl substituents that prohibits crystallization, is an amorphous polymer that slowly creeps under stress.

When propylene and ethylene are copolymerized, an amorphous copolymer (E–P) is formed at intermediate compositions. The subsequent cross-linking of these E–P copolymers results in a commercially important class of synthetic rubbers. E–P copolymers rich in ethylene are found to crystallize like the homopolymer polyethylene (PE), but with a greatly reduced content of long branches. These E–P copolymers are called linear, low-density PE, and are much easier to process than pure PE.

Amorphous polymers exhibit widely different bulk properties dependent on temperature. Below 100°C atactic polystyrene $[-(CH_2-CH-)_n]$ (PS) is rigid

and, as we all know, serves well as a material used to contain hot beverages in the form of styrofoam cups. At higher temperatures PS begins to lose its shape and to flow. The transition temperature between rigid (glassy) and soft

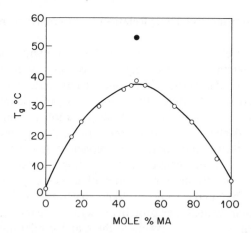

Figure 1.3 ■ Behavior of glass transition temperature as a function of comonomer composition for VDC–MA copolymers (Hirooka and Kato, 1974): ○, random; ●, regularly alternating. [Reprinted with permission from Tonelli (1975).]

(rubbery) behavior is termed the glass-to-rubber transition temperature, T_g (Ferry, 1970), and $T_g = 100°C$ for PS.

In Figure 1.3 the glass transition temperatures of a series of vinylidene chloride [$+CH_2-CCl_2+$]–methyl acrylate [$+CH_2-CH\underset{\underset{O=C-O-CH_3}{|}}{}+$]

(VDC–MA) copolymers are presented as a function of their overall comonomer composition. Clearly the T_g's of VDC–MA copolymers are elevated above the T_g's of their constituent homopolymers. Even more interesting is the observation that regularly alternating VDC–MA has a T_g considerably higher than a 50 : 50 VDC–MA copolymer with a random distribution of comonomer units. Both copolymers have the same overall composition, but differ in their microstructures, i.e. their comonomer sequence distributions. [See Tonelli (1975) for a possible explanation of this behavior.]

Having illustrated several among the myriad possible polymer microstructures and suggested how in a few instances they might influence the physical properties of polymers, we now begin to describe how nuclear-magnetic-resonance (NMR) spectroscopy can be used to unravel the microstructural details of polymers.

References

Carothers, W. H. (1931). *Chem. Rev.* **8**, 353.
Clark, H. G. (1968). *J. Polym. Sci. Part C* **16**, 3455.
Dworak, A., Freeman, W. J., and Harwood, H. J. (1985). *Polymer J.* **17**, 351.

Ferry, J. D. (1970). *Viscoelastic Properties of Polymers*, Second Ed., Wiley, New York.

Flory, P. J. (1953). *Principles of Polymer Science*, Cornell University Press, Ithaca, N.Y.

Flory, P. J. (1967). *J. Am. Chem. Soc.* **89**, 1798.

Hirooka, M. and Kato, T. (1974). *J. Polym. Sci. Polym. Lett. Ed.* **12**, 31.

Morawetz, H. (1985). *Polymers—The Origins and Growth of a Science*, Wiley-Interscience, New York.

Natta, G. (1955). *J. Am. Chem. Soc.* **77**, 1708.

Osgan, M. and Price, C. C. (1959). *J. Polym. Sci.* **34**, 153.

Price, C. C. and Osgan, M. (1956). *J. Am. Chem. Soc.* **78**, 4787.

Pruitt, M. E. and Baggett, J. M. (1955). U.S. Patent #2,706,181.

Shepherd, L., Chen, T. K., and Harwood, H. J. (1979). *Polymer Bull.* **1**, 445.

Staudinger, H. (1932). *Die Hochmolecularen Organischen Verbindungen*, Springer-Verlag, Berlin and New York.

Suter, U. W. and Neuenschwander, P. (1981). *Macromolecules* **14**, 528.

Tonelli, A. E. (1975). *Macromolecules* **8**, 544.

Tonelli, A. E. (1986). In *Cyclic Polymers*, J. A. Semlyen, Ed., Elsevier, London, Chapter 8.

Tsuruta, T. (1967). In *The Stereochemistry of Macromolecules*, A. D. Ketley, Ed., Marcel Dekker, New York.

Ward, I. M. (1985). *Adv. Polym. Sci.* **70**, 1.

Nuclear Magnetic Resonance

2.1. Introduction

The purpose of this chapter is to provide the reader with a brief introduction to the physics and measurement of the nuclear-magnetic-resonance (NMR) phenomenon. Most of the important concepts that will be used in later chapters to analyze polymer NMR spectra are merely introduced here. Consequently, the reader is encouraged to consult the general NMR texts referred to at the end of this chapter for a deeper understanding and appreciation of NMR spectroscopy.

While the nuclei of all atoms possess charge and mass, not every nucleus has angular momentum and a magnetic moment. Nuclei with odd mass numbers have spin angular-momentum quantum numbers I whose values are odd-integral multiples of $\frac{1}{2}$. Nuclei with even mass numbers are spinless if their nuclear charge is even, and have integral spin I if their nuclear charge is odd.

The angular momentum of a nucleus with spin I is simply $I(h/2\pi)$, where h is Planck's constant. If $I \neq 0$, the nucleus will possess a magnetic moment, μ, which is taken parallel to the angular-momentum vector. A set of magnetic quantum numbers m, given by the series

$$m = I, I-1, I-2, \ldots, -I, \qquad (2.1)$$

describes the values of the magnetic-moment vector which are permitted along any chosen axis. For nuclei of interest here (^{1}H, ^{13}C, ^{15}N, ^{19}F, ^{29}Si, ^{31}P), $I = \frac{1}{2}$, and thus $m = +\frac{1}{2}$ and $-\frac{1}{2}$. In general there are $2I + 1$ possible orientations of μ, or magnetic states of the nucleus. The ratio of the magnetic moment and the angular momentum is called the magnetogyric ratio, γ:

$$\gamma = 2\pi\mu/hI \qquad (2.2)$$

and is characteristic for a given nucleus.

The nuclei commonly observed in NMR studies of polymers usually have spin $I = \frac{1}{2}$, and, as we have seen, are characterized by $2I + 1 = 2$ magnetic states, $m = +\frac{1}{2}$ and $-\frac{1}{2}$. Both nuclear magnetic states have the same energy in the absence of a magnetic field, but they correspond to states of different potential energy upon application of a uniform magnetic field $\mathbf{H}_0$. The magnetic moment μ is either aligned along ($m = +\frac{1}{2}$) or against ($m = -\frac{1}{2}$) the field $\mathbf{H}_0$, with the latter state corresponding to a higher energy. Detection of the transitions of the magnetic nuclei between these spin states [$m = +\frac{1}{2}$ (parallel), $m = -\frac{1}{2}$ (antiparallel)] are made possible by the NMR phenomenon.

2.2. The NMR Phenomenon

2.2.1. Resonance

Let us discuss the interactions of magnetic fields applied to the magnetic moments of nuclei with spin $I = \frac{1}{2}$. In Figure 2.1 we have schematically drawn a nuclear magnetic moment μ in the presence of an applied magnetic field $\mathbf{H}_0$ acting along the z-axis of the coordinate system. The angle θ between the magnetic moment and the applied field does not change, because the torque,

$$\mathbf{L} = \mu \times \mathbf{H}_0, \tag{2.3}$$

tending to tip μ toward $\mathbf{H}_0$ is exactly balanced by the spinning of the magnetic moment, resulting in nuclear precession about the z-axis. Increasing H_0 in an attempt to force the alignment of μ along the z-axis only results in faster precession. A good analogy is provided by the precession of a spinning top in the earth's gravitational field.

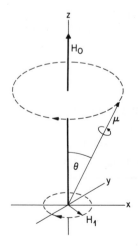

Figure 2.1 ■ Nuclear magnetic moment in a magnetic field.

The precessional or Larmor frequency, ν_0, of the spinning nucleus is given by

$$\nu_0 = \frac{\gamma}{2\pi} H_0, \tag{2.4}$$

and is independent of θ. However, the energy of the spin system does depend on the angle between μ and H_0:

$$E = -\mu \cdot H_0 = -\mu H_0 \cos\theta. \tag{2.5}$$

We may change the orientation θ between μ and H_0 by application of a small rotating magnetic field H_1 orthogonal to H_0 (see Figure 2.1). Now μ will experience the combined effects of H_1 and H_0 if the angular frequency of H_1 coincides with ν_0, the precessional frequency of the spin. The nucleus absorbs energy from H_1 in this situation and θ changes; otherwise H_1 and μ would not remain in phase and no energy would be transferred between them.

If the rotation rate of H_1 is varied through the Larmor frequency of the nucleus, a resonance condition is achieved, accompanied by a transfer of energy from H_1 to the spinning nucleus and an oscillation of the angle θ between H_0 and μ. At $H_0 = 2.34$ T (1 T = 1 tesla = 10 kilogauss) the resonant frequencies of the ^{1}H, ^{19}F, ^{31}P, ^{13}C, ^{29}Si, and ^{15}N nuclei are $\nu_0 = 100$, 94, 40.5, 25.1, 19.9, and 10.1 MHz, respectively.

2.2.2. Interactions and Relaxations of Nuclear Spins

Figure 2.2 illustrates the magnetic energy levels for a spin-$\frac{1}{2}$ nucleus in a magnetic field H_0. The energy separation between nuclear spin states is

$$\Delta E = 2\mu H_0, \tag{2.6}$$

and the relative populations of the upper $(+)$ and lower $(-)$ states is given by the Boltzmann expression

$$\frac{N_+}{N_-} = \exp\left(-\frac{\Delta E}{kT}\right) = \exp\left(-\frac{2\mu H_0}{kT}\right). \tag{2.7}$$

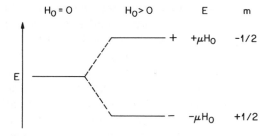

Figure 2.2 ■ Energy levels for a spin-$\frac{1}{2}$ nucleus in a magnetic field H_0.

The excess population of the lower energy state is

$$\frac{N_- - N_+}{N_-} = \frac{2\mu H_0}{kT}, \tag{2.8}$$

where the approximation $e^{-X} = 1 - x$, for small x, has been adopted.

At a field strength $H_0 = 2.34$ T the separation between magnetic energy levels for proton nuclei is $\sim 10^{-2}$ cal, which results in an excess population of $\sim 2 \times 10^{-5}$ spins of lower energy. For an assemblage of nuclei, this small spin population difference leads to a correspondingly small macroscopic moment directed along H_0. Removal of H_0 results in a loss of the macroscopic moment, because the magnetic energy levels are degenerate in the absence of the field. By what mechanism(s) is the Boltzmann distribution of spin states established following the application of H_0, and how long does it take?

An alternative way of phrasing this question might be, what mechanisms are responsible for relaxing upper-level spins to their lower level after application of H_0, thereby maintaining parity between the spin and sample temperatures? Such a relaxation is made possible because each spin is not completely isolated from the rest of the molecules in the sample, called the *lattice*. The spins and the lattice may be considered to be largely separate coexisting systems which are weakly coupled through an inefficient yet very important link by which thermal energy may be exchanged. The molecular motions of the neighboring nuclei which constitute the lattice provide the mechanism for transferring thermal energy between the spins and their surroundings.

The relative motions of neighboring nuclei generate fluctuating magnetic fields which are experienced by the observed nucleus as it precesses about the direction of the applied field H_0. A broad range of frequencies will be associated with the fluctuating fields produced by the lattice motions, because these motions, relative to the observed nucleus, are nearly random. Components of the fluctuating magnetic fields generated by the lattice motions which lie along H_1 (see Figure 2.1) and have frequency ν_0 will, like H_1, induce transitions between the magnetic energy levels of the observed nuclei. The rates of this spin–lattice relaxation must therefore be directly connected to the rates of molecular motions in the lattice.

The spin–lattice relaxation time, T_1, is the time required for the difference between the excess and equilibrium spin populations to be relaxed by a factor of e. For liquids T_1 is usually in the range 10^{-2}–10^2 sec, while in solid samples T_1 may be as long as hours. Spin–lattice relaxation produces a change in energy via a redistribution of magnetic moments with components along the applied field H_0. As a result, T_1 is often termed the longitudinal relaxation time: it is associated with a decay of the macroscopic nuclear moment along the direction of the applied field H_0 (the z-direction; see Figure 2.1).

There is a second mode by which nuclear magnetic moments may interact. This interaction is illustrated in Figure 2.3. Here a pair of nuclear moments

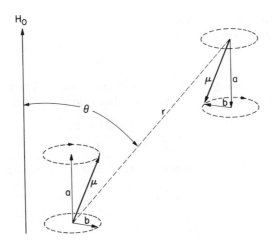

Figure 2.3 ▪ A pair of precessing nuclear moments with static (**a**) and rotating (**b**) components indicated.

are precessing about the H_0-axis, and each is decomposed into a static component along H_0 (**a**) and a rotating component in the xy plane (**b**). If the rotating component precesses at the Larmor frequency ν_0, then a neighboring nucleus may be induced to undergo a spin transition, resulting in a spin exchange. No net change in the total energy is produced by the exchange of neighboring nuclear spins, but clearly the lifetimes of the interacting spins are affected. This exchange of neighboring nuclear spins is called spin–spin relaxation and is characterized by T_2, the spin–spin relaxation time. T_2 is also called the transverse relaxation time, because it is concerned with the rate of change of magnetization in the xy plane, which is transverse to the H_0 field direction.

2.2.3. Chemical Shift

We have seen that by application of a rotating magnetic field H_1 transverse to the static field H_0, about which a spinning nuclear magnet is precessing, we can flip the nuclear spin by rotating H_1 at the precession or Larmor frequency ν_0. If all nuclei of the same type, e.g. all protons, were to resonate at the same field strength H_0, then NMR would not be a spectroscopic tool useful for the study of molecular structure. Fortunately, soon after the demonstration of NMR in condensed phases, it was observed that the characteristic resonant frequency of a nucleus depends upon its chemical or structural environment.

The cloud of electrons about each nucleus produces orbital currents when placed in a magnetic field H_0. These currents produce small local magnetic fields which are proportional to H_0 but are opposite in direction, thereby effectively shielding the nucleus from H_0. Consequently, a slightly higher value of H_0 is needed to achieve resonance. The actual local field H_{loc} experienced by a nucleus can be expressed as

$$H_{loc} = H_0(1 - \sigma), \tag{2.9}$$

where σ is the screening constant. σ is highly sensitive to chemical structure but independent of H_0. The resonant Larmor frequency becomes

$$\nu_0 = \frac{\mu H_{loc}}{hI} = \frac{\mu H_0(1 - \sigma)}{hI}, \tag{2.10}$$

and the separation between magnetic energy levels is now (see Eq. 2.6)

$$\Delta E = 2\mu H_{loc} = 2\mu H_0(1 - \sigma). \tag{2.11}$$

Nuclear screening decreases the spacing of nuclear magnetic energy levels. An increase in the magnetic shielding requires an increase in H_0 at constant ν_0 and a decrease in ν_0 at constant field strength H_0 to achieve resonance.

Nuclear shielding is influenced by the number and types of atoms and groups attached to or near the observed nucleus. The dependence of σ upon molecular structure lies at the heart of NMR's utility as a probe of molecular structure. We will have more to say about the structural dependence of σ, i.e. the chemical shift, when we discuss the ^{13}C NMR of polymers in Chapter 4.

There is no natural fundamental scale unit in NMR spectroscopy. Both the energies of transition between spin quantum levels and the nuclear shielding produced by the screening constant σ are proportional to the applied field H_0. In addition, there is no natural zero of reference in NMR. These difficulties are avoided by expressing the resonant frequencies of nuclei in parts-per-million (ppm) relative changes in H_0 and referring the observed changes or displacements in resonance, called chemical shifts, to the ppm relative change in the resonant frequency of an arbitrary reference substance added to the sample. In ^{1}H and ^{13}C NMR spectroscopy, for example, it is customary to use tetramethylsilane (TMS) as the reference compound, where the chemical shifts δ of both the ^{1}H and ^{13}C nuclei are taken as $\delta = 0$ ppm.

2.2.4. Spin–Spin Coupling

We close this discussion of the physics of the NMR phenomenon by describing two important modes of nuclear spin–spin coupling: direct (through space) dipolar coupling and indirect (through intervening chemical bonds) scalar coupling. Two neighboring nuclear spins will feel, in addition to $\mathbf{H}_0$, the local magnetic field $\mathbf{H}_{loc}$ produced by each other. H_{loc} is given by

$$H_{loc} = \pm \mu r^{-3}(3\cos^2\theta - 1), \tag{2.12}$$

where r is the distance between nuclei and θ the angle between $\mathbf{H}_0$ and the line joining the nuclei (see Figure 2.3). The fact that H_{loc} may add to or subtract from H_0 depending upon whether the neighboring magnetic dipole is aligned with or against $\mathbf{H}_0$ is reflected by the $\pm$ sign. This form of spin–spin coupling is called dipolar coupling and serves to broaden the resonance line of a nucleus.

There are two important situations where dipolar coupling does not contribute to line broadening. The first is when all neighboring nuclei are rigidly oriented at the magic angle of $\theta = 54.7°$, where $\cos^2\theta = \frac{1}{3}$ and $H_{\mathrm{loc}} = 0$ (see Eq. 2.12). If the relative orientations of neighboring spins vary rapidly with respect to the time a nucleus spends in a given spin state, i.e. with respect to the spin–spin relaxation time T_2, then H_{loc} is given by its space average

$$H_{\mathrm{loc}} = \mu r^{-3} \int_0^{\pi} (3\cos^2\theta - 1)\sin\theta \, d\theta, \qquad (2.13)$$

which also vanishes. Both of these circumstances are important for observing high-resolution NMR spectra of polymers and will be discussed further in subsequent chapters.

Nuclear spins may also be coupled by orbital motions of their valence electrons or polarization of their spins occurring indirectly through the intervening chemical bonds. Unlike dipolar coupling of nuclear spins, this indirect or scalar coupling is not affected by molecular tumbling and is also independent of H_0. Two spin-$\frac{1}{2}$ nuclei so coupled will each split the other's resonance into a doublet, because in a large collection of such nuclear pairs the probabilities of each finding the other's spin with $(+\frac{1}{2})$ or against $(-\frac{1}{2})$ $\mathbf{H}_0$ are nearly equal. If one nucleus of the pair is further coupled to a second group of two identical nuclei with $++$, $+-$ $(-+)$, and $--$ spin orientations, then the resonance of the first nucleus will appear as a $1:2:1$ triplet. The resonance of the identical pair will be a doublet. A single nucleus coupled to three equivalent neighboring spins with $+++$; $++-$, $+-+$, $-++$ $(+--, -+-, --+)$; and $---$ orientations would exhibit a $1:3:3:1$ quartet resonance. A spin-$\frac{1}{2}$ nucleus with n equivalently coupled neighbors also of spin $\frac{1}{2}$ will have its resonance split into $n+1$ peaks.

In the NMR spectra of polymers only $^1\mathrm{H}-^1\mathrm{H}$, $^{13}\mathrm{C}-^1\mathrm{H}$, $^{13}\mathrm{C}-^{19}\mathrm{F}$, $^{15}\mathrm{N}-^1\mathrm{H}$, $^{19}\mathrm{F}-^{19}\mathrm{F}$, $^{19}\mathrm{F}-^1\mathrm{H}$, $^{29}\mathrm{Si}-^1\mathrm{H}$, and $^{31}\mathrm{P}-^1\mathrm{H}$ scalar couplings are important. The magnitude and sign of the scalar coupling of two magnetic nuclei depend on substituents and geometry. The strength of the coupling in hertz is designated ^xJ, where the superscript x denotes the number of intervening chemical bonds between the coupled nuclei. A particularly useful relation is based on the observed geometry-dependent vicinal $^1\mathrm{H}-^1\mathrm{H}$ coupling ^3J illustrated in Figure 2.4. Here it is observed that when the vicinal protons are *trans* (a) the scalar coupling is large (~ 12 Hz), but it is markedly reduced (~ 2 Hz) in their *gauche* arrangement (b). An example illustrating how the conformationally sensitive ^3J coupling between vicinal protons may be used to study the conformation and microstructure of a polymer is presented in Chapter 6.

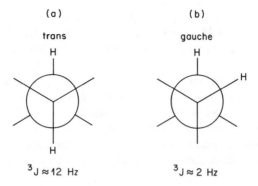

(a)   (b)

trans   gauche

$^3J \approx 12$ Hz $^3J \approx 2$ Hz

Figure 2.4 ■ Vicinal ^3J ^{1}H–^{1}H scalar coupling in a saturated hydrocarbon.

2.3. Experimental Observation of NMR

In Figure 2.5 we present a drawing of a superconducting magnet used in a modern high-field NMR spectrometer. The magnet is bathed in liquid helium to maintain its superconductivity. The radio frequency coil provides the rf energy appropriate to excite the nuclei in the sample to resonance.

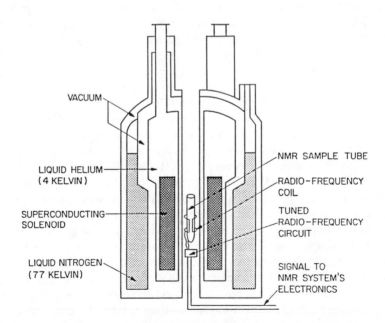

VACUUM

NMR SAMPLE TUBE

LIQUID HELIUM
(4 KELVIN)

RADIO–FREQUENCY
COIL

SUPERCONDUCTING
SOLENOID

TUNED
RADIO–FREQUENCY
CIRCUIT

LIQUID NITROGEN
(77 KELVIN)

SIGNAL TO
NMR SYSTEM'S
ELECTRONICS

Figure 2.5 ■ Cross section of a superconducting NMR magnet. Magnet assembly has a diameter of 70 cm, while the sample tube is 1 cm in diameter. [Reprinted with permission from Bovey and Jelinski (1987).]

The degeneracy of the nuclear magnetic spin energy levels is removed by the static magnetic field H_0. Application of the rotating magnetic, or electromagnetic, field H_1 excites transitions between these energy levels. When the frequency of the H_1 field (radio frequency, or rf, in megahertz) is equal to the Larmor frequency of the observed nucleus, the resonance condition occurs, i.e. when

$$H_1 = \nu_0 = \gamma \frac{H_0}{2\pi}. \tag{2.14}$$

Most samples will have multiple Larmor frequencies, because most molecules have more than a single magnetically equivalent group (CH, CH_2, CH_3 for example), leading to several resonance frequencies or chemical shifts. The method used to excite the nuclei and achieve resonance must clearly be capable of covering all of the Larmor frequencies in the sample.

Two principal methods have been developed to achieve the resonance condition in NMR spectroscopy—continuous wave (CW) and Fourier transform (FT). In the CW method each magnetically equivalent nucleus is successively made to resonate by sweeping either of the magnetic fields, the rf H_1 or the static H_0 field. As each nucleus is brought into resonance by the field-sweeping process, a voltage is induced in the rf pickup coil (see Figure 2.5).

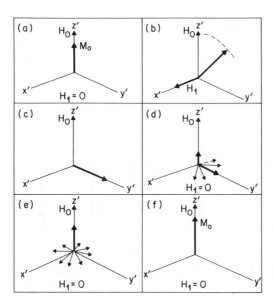

Figure 2.6 ■ Pulsed NMR experiment in the rotating frame. (a) Net magnetization $\mathbf{M}_0$ aligned along $\mathbf{H}_0$; (b,c) rf field $\mathbf{H}_1$ applied perpendicular to $\mathbf{H}_0$ for a duration sufficient to tip $\mathbf{M}_0$ by 90° into the $x'y'$ plane; (d,e) spins begin to relax in the $x'y'$ plane by spin–spin (T_2) processes, and in the z'-direction by spin–lattice (T_1) processes; (f) equilibrium $\mathbf{M}_0$ is reestablished along $\mathbf{H}_0$.

After amplification, this signal is detected directly in the frequency domain and recorded in a plot of voltage (intensity) versus frequency.

By contrast, the FT method employs signal detection in the time domain followed by a Fourier transformation into the frequency domain. Simultaneous excitation of all the Larmor frequencies is produced by applying a pulse (short burst) of rf signal at or near ν_0, resulting in an equalization of the populations of the nuclear spin energy levels. Equilibrium spin populations are reestablished in a free-induction-decay (FID) process following the rf pulse. The vector diagram in Figure 2.6 can be used to visualize the effect of the rf pulse (H_1) on the nuclear spins and their subsequent FID to equilibrium.

At equilibrium in the presence of H_0, more spins will be aligned along H_0 then against it, and this is indicated by the net magnetic moment M_0 drawn along the field direction z' in Figure 2.6(a). (Note that the primes indicate that the reference frame $x'y'z'$ is rotating at the Larmor frequency.) The net magnetization has been tipped 90° into the $x'y'$ plane [Figure 2.6(b,c)] by application of an rf pulse H_1 whose duration is just sufficient to equalize the magnetic energy levels, i.e., $M_0 = 0$ along z'.

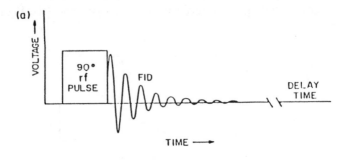

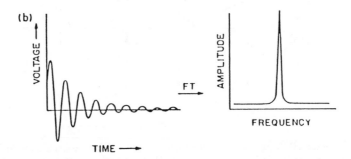

Figure 2.7 ■ (a) Repetitive rf pulse sequence. (b) Fourier transformation of the time-domain FID into the frequency-domain NMR signal.

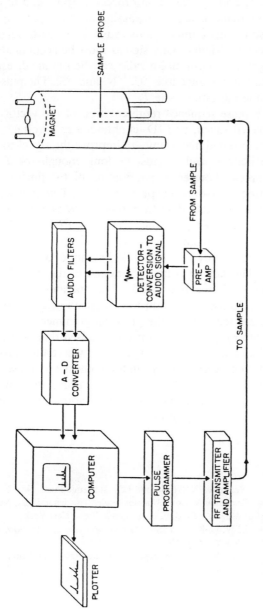

Figure 2.8 ■ Block diagram of a pulsed FT NMR spectrometer. [Reprinted with permission from Bovey and Jelinski (1987).]

Following the rf pulse [Figure 2.6(d, e)], the spins begin to reestablish their initial state through T_1 and T_2 relaxation processes. Spin–spin interactions in the transverse $(x'y')$ plane cause a dephasing of the spins in this plane (T_2-process), while spin–lattice interactions cause the spins to relax along the z'-direction (T_1-process). Usually many signals must be accumulated before a spectrum with adequate signal-to-noise ratio can be obtained, especially for nuclei with low natural abundance like ^{13}C, ^{15}N, and ^{29}Si. The pulse repetition rate is governed by the relaxation time T_1.

Figure 2.7 gives a pulse sequence representation of the vector diagram in Figure 2.6. The detected signal, or FID, is obtained as a voltage in the time domain. The pulse is repeated many times to improve the signal-to-noise ratio, and the delay time between pulses must be long enough for T_1 relaxation processes to be complete. Fourier transformation of the time-domain signal results in the usual frequency-domain spectrum. The FT method saves time by collecting data all at once, rather than from a slow CW sweep of the field, and is well suited to signal averaging by collecting many FIDs from weak signals before Fourier-transforming them.

Finally, in Figure 2.8 we present the block diagram of a modern pulsed FT NMR spectrometer. The sample is first positioned in the most homogeneous part of the magnetic field (see Figure 2.5). The operator then instructs the computer to signal the pulse programmer to begin the experiment. Precisely timed digital pulses are sent out by the pulse programmer, and rf pulses are generated by superimposing rf signals on these pulsed digital signals. The rf pulses are amplified and sent to the sample probe, where they produce FIDs. Upon amplification and detection by audio conversion, these signals are filtered and converted to a digital representation using an analog-to-digital (A–D) converter. These digital signals are finally stored in the computer for further processing and eventual plotting.

References

Becker, E. D. (1980). *High Resolution NMR*, Second Ed., Academic Press, New York.

Bovey, F. A. (1972). *High Resolution NMR of Macromolecules*, Academic Press, New York.

Bovey, F. A. (1988) *Nuclear Magnetic Resonance*, Academic Press, New York.

Bovey, F. A. and Jelinski, L. W. (1987). *Nuclear Magnetic Resonance*, Encyclopedia of Polymer Science, Vol. 10, Wiley, New York, p. 254.

Stothers, J. B. (1972). *Carbon-13 NMR Spectroscopy*, Academic Press, New York.

3

High-Resolution NMR
of Polymers

3.1. Introduction

Though NMR spectroscopy is over forty years old, it still remains in a state of rapid development. Magnetic field strengths of current superconducting NMR spectrometers are nearly 40-fold greater than those employed in the first permanent-magnet prototype spectrometers. Pulse programmable FT spectrometers permit the selective observation of nuclei based on their unique structural (chemical) and motional characteristics. Almost daily new concepts (2D NMR, cross-polarization, etc.) and new techniques (INEPT, DEPT, magic-angle sample spinning, etc.) are reported and applied to a variety of molecular systems including both synthetic and biopolymers.

Though the first report of polymer NMR spectra, a wide-line ^{1}H NMR study of natural rubber (Alpert, 1947), appeared the year after the discovery of the NMR phenomenon in bulk matter (Purcell et al., 1946; Bloch et al., 1946), it was not until the late 1950s that high-resolution NMR spectra were observed for polymers. Even from polymer solutions, which are often very viscous, reasonably well-resolved NMR spectra can be obtained, such as those reported for ribonuclease (Saunders et al., 1957) and polystyrene (Saunders and Wishnia, 1958; Odajima, 1959; Bovey et al., 1959).

It is the rapid local motions of polymer chain segments (nanosecond to picosecond range) which produce high-resolution NMR spectra for dissolved polymers. The volumes pervaded by dissolved macromolecules are much larger than their molecular volumes and produce solutions of high viscosity through polymer–polymer entanglements and entrapment of surrounding solvent molecules. However, as we have seen in the previous chapter (see Sections 2.2.2–4), both the frequency at which a magnetic nucleus resonates and the width of the resulting resonance peak depend on the local structure and its motional dynamics in the immediate vicinity of the observed nucleus. Thus NMR serves as a local microscopic probe of molecular structure and its motions, and can even provide highly resolved spectra for dissolved polymers, whose overall motion may be sluggish, but whose local segmental motions are rapid.

3.2. ^{1}H NMR

Historically the proton (^{1}H) was the first nucleus observed in the magnetic-resonance spectroscopy of polymers. The 500-MHz proton NMR spectra recorded on a superconducting magnet (11.7 T) for two samples of poly(methyl methacrylate) (PMMA) are presented in Figure 3.1 (Schilling et al., 1985). A free-radical initiator was used in the polymerization of the predominantly syndiotactic sample (s-PPMA) in (a), while the isotactic sample (i-PPMA) in (b) was obtained by anionic initiation. It is apparent from the methylene proton portion of both spectra that free-radical and anionic initiated polymerization of methyl methacrylate results in PMMA samples with very different microstructures.

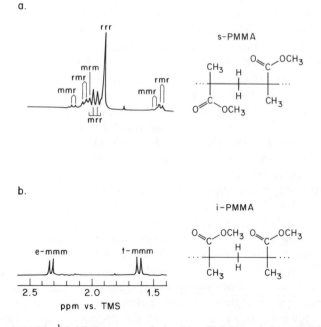

Figure 3.1 ■ 500-MHz ^{1}H NMR spectra of the (a) s-PMMA and (b) i-PMMA. Only the methylene proton regions are shown. [Reprinted with permission from Schilling et al. (1985).]

The methylene protons in the racemic (r) diad drawn in Figure 3.1(a) are magnetically equivalent, because of the twofold axis of symmetry present in an r-diad. They resonate at the same frequency, leading to a singlet, despite the strong two-bond, geminal ^2J coupling between them. In the meso (m) diad of Figure 3.1(b), which lacks an axis of symmetry, the methylene protons are magnetically nonequivalent and appear as a pair of doublets, each with a spacing of ~ 15 Hz produced by their ^2J geminal coupling.

The PMMA sample produced by anionic initiation does show (b) almost exclusively a pair of doublets, indicating that nearly all of its diads are m, or that this PMMA is isotactic. The principal methylene proton resonance observed for the free radical PMMA is a singlet at 1.9 ppm [see (a)], meaning that most of its diads are r and that this sample is predominantly syndiotactic, though more irregular than the anionically initiated sample. It is apparent from this example that ^1H NMR can provide absolute stereochemical information about a vinyl polymer without recourse to other methods, such as x-ray diffraction.

The 220-MHz ^1H NMR spectra of three polypropylene (PP) samples are presented in Figure 3.2 (Ferguson, 1967a, b; Heatly and Zambelli, 1969). Note the apparent greater resolution of the spectra in (a) and (c) recorded for the stereoregular samples (isotactic and syndiotactic) than for atactic PP in (b). The impression of degraded resolution in the spectrum of atactic PP is a consequence of the overlapping of many slightly different chemical shifts corresponding to the various triad and tetrad stereosequences present in the atactic sample. Only the rr (rrr) and mm (mmm) triads (tetrads) are present in the syndiotactic and isotactic samples, respectively.

In addition to the geminal coupling (^2J) of methylene protons in meso diads, the methine and methyl and the methylene and methine protons show significant vicinal, three-bond couplings (^3J) in nearly all stereosequences. By the technique of ^1H–^1H homonuclear decoupling, or double resonance, some of these couplings can be removed. The scalar, or J, coupling between two magnetically nonequivalent nuclei A, B can be removed by irradiating B with a strong rf field H_2 tuned to its resonance frequency while observing A with the weaker H_1 field. H_2 causes such a rapid oscillation of B between its spin states that it no longer couples to A. Unfortunately, in PP the methine protons are coupled to both the methylene and methyl protons, requiring a triple resonance experiment to simultaneously remove all vicinal couplings. Because the methylene protons in m-diads resonate at widely separated frequencies, which in turn are also distinct from the methylene protons in r-diads (see Figure 3.2), complete removal of the vicinal couplings observed in the ^1H NMR spectra of PP seems unlikely.

In the next section, where we begin to discuss the ^{13}C NMR of polymers, it will become apparent why ^1H NMR has been supplanted by ^{13}C NMR in the study of polymer microstructure. ^{13}C NMR usually results in spectra free of the extensive overlapping of resonances belonging to different stereosequences,

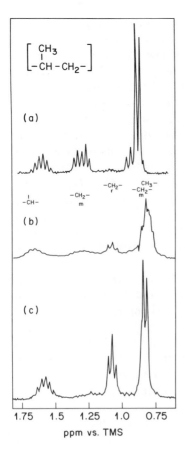

1.75   1.5   1.25   1.0   0.75
ppm vs. TMS

Figure 3.2 ■ 220-MHz ^{1}H NMR spectra of (a) isotactic, (b) atactic, and (c) syndiotactic polypropylene.

as occurs in the ^{1}H NMR spectra of polymers like those shown for PP in Figure 3.2.

3.3. ^{13}C NMR

The ^{13}C nucleus occurs at a natural abundance of only 1.1% and has a small magnetic moment—about one-fourth that of the proton. Both factors tend to mitigate against the observation of high-resolution ^{13}C NMR spectra. However, the decrease in observational sensitivity of the ^{13}C nucleus can be compensated for by employing the pulsed FT technique, combined with spectrum accumulation, as described in Section 2.3. The time saved by the pulsed FT recording of spectra makes possible the accumulation of a sufficient number of spectra to produce a suitable signal-to-noise ratio.

Further increase in ^{13}C signal intensity is obtained by removing the nuclear-spin coupling between ^{13}C nuclei and their directly bonded protons and from the accompanying nuclear Overhauser enhancement (Stothers, 1972). Removal of the strong (125–250 Hz) ^{13}C–^{1}H nuclear coupling, by providing a second rf field at the proton resonance frequency, results in the collapse of ^{13}C multiplets and an improved signal-to-noise ratio. Saturation of nearby protons produces a nonequilibrium polarization of the ^{13}C nuclei, which exceeds the thermal value, and yields an increase in the observed signal strength. It has been demonstrated (Kuhlmann and Grant, 1968) that the dipole–dipole coupling mechanism dominates for the ^{13}C isotope and a maximum nuclear Overhauser enhancement (NOE) factor of 3 is produced by a directly bonded proton.

Another means of increasing the sensitivity of ^{13}C NMR spectroscopy can be realized by transferring the polarization of a sensitive nucleus say ^{1}H (large γ) to the insensitive nucleus ^{13}C (small γ). This is achieved by the technique of selective population transfer (SPT) (Derome, 1987) and can enhance the ^{13}C signal intensity by a factor of $\gamma_H/\gamma_C = 4$. By application of the appropriate pulse sequences (INEPT and DEPT), not only can the ^{13}C signals be enhanced, but they can be edited to differentiate between CH, CH_2, and CH_3

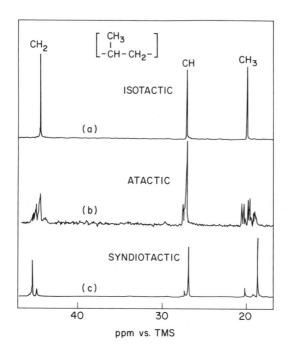

Figure 3.3 ■ 25-MHz ^{13}C NMR spectra of (a) isotactic, (b) atactic, and (c) syndiotactic PP. [Reprinted with permission from Tonelli and Schilling (1981).]

resonances. An application of the INEPT and DEPT pulse editing techniques is discussed in Chapter 7.

Having discussed several of the means utilized to overcome the inherent insensitivity of the ^{13}C nucleus, let us now mention the principal advantage of ^{13}C NMR spectroscopy of organic molecules, including polymers. It is the increased sensitivity of ^{13}C shieldings to molecular structure, in the 200-ppm range for neutral organics compared to 10–12 ppm for ^{1}H shieldings, which has resulted in the replacement of ^{1}H NMR by ^{13}C NMR as the method of choice in molecular structure investigations.

The superior sensitivity of ^{13}C NMR to molecular structure is easily demonstrated by comparing the 25-MHz ^{13}C NMR spectra observed for three PP samples in Figure 3.3 (Tonelli and Schilling, 1981) with the corresponding ^{1}H NMR spectra shown in Figure 3.2. The ^{13}C resonances are spread over an ~ 30-ppm range compared to the < 1-ppm range exhibited by the ^{1}H resonances. Unlike homonuclear ^{1}H–^{1}H coupling, ^{13}C–^{1}H heteronuclear coupling is always easily removed, leading to spectral simplification. Both of these advantages result in the kind of sensitivity to microstructure seen in the methyl regions of the ^{13}C NMR spectra of atactic and isotactic PP presented in Figure 3.4.

Here we can distinctly observe resonances for nearly all 10 possible pentad stereosequences. The assignment of resonances to individual stereosequences,

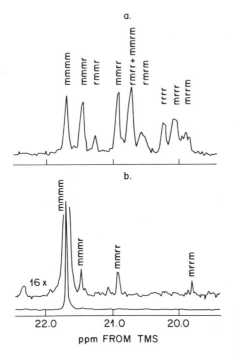

ppm FROM TMS

Figure 3.4 ■ Expanded methyl regions of the 25-MHz ^{13}C NMR spectra of atactic (a) and isotactic (b) PP.

aside from the obvious *mmmm* isotactic resonance [compare Figure 3.4(a) and (b)], will be described in greater detail in Chapter 6:

PP – PENTADS

As we have already mentioned, long polymer chains move segmentally, at least on a local level. Therefore, the spin–lattice relaxation of their nuclei will not be proportional to the macroscopic solution viscosity, but will instead depend on the microscopic solvent viscosity. The line widths observed in solution, which reflect local chain motion, will then depend on the viscosity of the solvent, and not on the overall polymer solution viscosity. Clearly solvents of moderate to low viscosities are to be preferred.

However, in order to reduce dipolar broadening it is usually advantageous to record polymer spectra at high temperatures (T > 100°C), where the polymer segments move rapidly. At these high temperatures solvents of low viscosity, which are highly volatile, will boil. In addition, semicrystalline polymers are often difficult to dissolve unless heated in the vicinity of their melting points, because of the small entropy of mixing associated with the dissolution of long-chain macromolecules (Flory, 1954; Morawetz, 1975). The best compromise for polymer NMR is a high-boiling solvent that dissolves the polymer in question and does not itself resonate in the same frequency ranges as the polymer. For most vinyl polymers the chlorinated benzenes make useful NMR solvents.

TMS is normally not useful as a reference substance for polymer NMR spectroscopy, because it evaporates at the elevated temperatures usually employed. Hexamethyldisiloxane (HMDS), which is much less volatile, is most often used as a chemical-shift reference in polymer NMR. The carbon nuclei in HMDS resonate 2 ppm downfield from TMS.

To insure quantitative ^{13}C NMR spectra, the rf H_1 field pulses should be sufficiently separated so that full spin–lattice relaxation is realized for all ^{13}C nuclei in the sample. If the repetition rate of the rf pulses approaches the T_1 of some of the ^{13}C nuclei in the sample, then incorrect relative signal intensities will result. As a practical, operational guide the delay between rf pulses should be five times the T_1 of the slowest-relaxing nucleus in the sample (Farrar and Becker, 1971). All of the spectra discussed quantitatively in later chapters have been recorded in this manner.

3.4. High-Resolution ^{13}C NMR in the Solid State

3.4.1. Dipolar Broadening

In Chapter 2 (Section 2.2.4) we mentioned the direct coupling of neighboring nuclear spins through space as a consequence of the local magnetic fields produced by each other. For dilute spins like ^{13}C ($\sim 1\%$ natural abundance), it is the abundant ^{1}H spins which produce the local fields at the ^{13}C nuclei. The local field H_{loc} produced at a ^{13}C nucleus by a proton dipole is given by

$$H_{loc} = \pm \frac{h\gamma_H}{4\pi} \frac{3\cos^2\theta - 1}{r^3}. \tag{3.1}$$

This local field may add to or subtract from the applied field H_0, depending on whether or not the proton dipole is aligned with or against $\mathbf{H}_0$, and thus the $\pm$ sign. The internuclear vector $\mathbf{r}$ and the angle θ are illustrated and defined in Figure 2.3. If all ^{13}C nuclei in a given sample had neighboring protons fixed at the same r and θ, then the ^{13}C resonance would be split into two components whose separation would depend on the sample's orientation in the magnetic field.

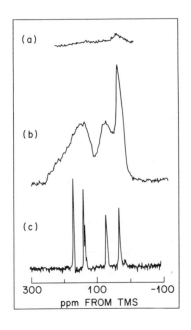

Figure 3.5 ■ ^{13}C NMR spectra of bulk PBT obtained using lower-power decoupling (a), high-power (dipolar) decoupling (b), and dipolar decoupling with rapid magic-angle sample spinning (c). [Reprinted with permission from Jelinski (1982).]

In rigid polymer samples, which are glasses or microcrystalline, a whole range of r and θ values occur, leading to a broad distribution of local fields. Their summation results in a dipolar broadening of many kHz, which is more than enough to mask all chemical-shift, and consequently microstructural, information. The rate of molecular segmental motion in a polymer solution is sufficient to produce a sampling of all dipolar orientations θ in a time short compared to the dipolar coupling. This results in a small dipolar broadening, because the time average of $3\cos^2\theta - 1$ (see Eq. 3.1) can be replaced by its space average, which vanishes.

We may remove the broadening of solid-state ^{13}C resonances produced by their dipolar coupling to neighboring ^1H nuclei in the same manner that scalar J-couplings are removed in solution (see Section 3.2). However, the strength of the auxiliary applied field H_2 must now be of kilohertz magnitude, and not hertz as in scalar decoupling. In Figure 3.5(a) and (b) (Jelinski, 1982) we compare the ^{13}C NMR spectra of solid poly(butylene terephthalate) (PBT) obtained with and without high-power proton decoupling. A decoupling field H_2 of about 50 kHz is necessary to drive or flip the ^1H spins at a rate which is rapid compared to the static ^{13}C–^1H dipole–dipole interaction. Though a significant improvement in resolution has resulted from the high-power ^1H decoupling, the spectrum in Figure 3.5(b) falls far short of the high-resolution ^{13}C NMR spectra recorded in solution. The remaining line broadening is due primarily to chemical-shift anisotropy.

3.4.2. Chemical-Shift Anisotropy

In Chapter 2 (Section 2.2.3) we discussed the fact that nuclei in different electronic environments resonate at different frequencies, or have different chemical shifts δ, because they are shielded from the applied field $\mathbf{H}_0$ to different extents. Throughout this book we utilize the local structural dependence of the chemical shift to characterize the microstructures and conformations of polymers. The external applied field $\mathbf{H}_0$ produces electronic currents in a molecule, and these currents in turn produce a local magnetic field at the nucleus.

This three-dimensional local field can be described in terms of its magnitude and molecular orientation by the chemical-shift tensor σ, given by

$$\sigma = \sigma_{11}\lambda_{11}^2 + \sigma_{22}\lambda_{22}^2 + \sigma_{33}\lambda_{33}^2. \tag{3.2}$$

The σ's, or principal values of the chemical-shift tensor, give the magnitude of the tensor in three mutually perpendicular directions (Cartesian coordinates), and the λ's are the direction cosines which specify the orientation of the molecular principal coordinate system with respect to the applied field. In solution the isotropic chemical shift, σ_i, is observed, due to the rapid molecular motion which averages σ over all orientations:

$$\sigma_i = \tfrac{1}{3}(\sigma_{11} + \sigma_{22} + \sigma_{33}) = \tfrac{1}{3}\,\text{trace }\sigma. \tag{3.3}$$

It is apparent from Eq. 3.2 that in a rigid solid sample the chemical shift of a particular nucleus will depend on its orientation with respect to the applied field. A sample having all carbon nuclei with the same orientation, as in a single crystal, will exhibit a chemical shift that varies as the crystal is rotated in the field. In a powdered sample all possible crystalline orientations are present, and the NMR spectrum will consist of the chemical-shift-tensor powder pattern.

Two theoretical chemical-shift-tensor powder patterns are illustrated in Figure 3.6. Principal values σ_{11}, σ_{22}, and σ_{33} are indicated, and their isotropic averages, σ_i, are given as dotted lines. $\sigma_{\parallel}$ and $\sigma_{\perp}$ in the axially symmetric case correspond to the resonant frequencies observed when the principal-axis system is parallel and perpendicular, respectively, to the applied field. Molecular motion will narrow the chemical-shift tensor, and the resulting powder pattern contains information concerning the axis and angular range of the motion.

Though potentially capable of providing motional information, the chemical-shift-tensor powder pattern contributes significantly to the broadening of solid-state NMR spectra and often obscures the structural information available from the isotropic chemical shifts. The broadening of resonances in the

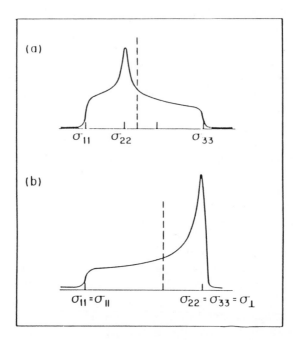

Figure 3.6 ■ Schematic chemical-shift-tensor powder pattern for an axially asymmetric (a) and axially symmetric (b) tensor. The isotropic chemical-shift values σ_i are indicated as dashed lines.

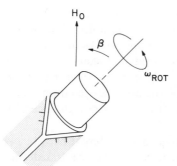

Figure 3.7 ■ Typical Andrews (1959) design sample holder (rotor) rotating on air bearings within a stator (shaded).

solid state may be removed by the high-speed spinning of the sample at the magic angle.

If the sample is rotated rapidly about an axis at an angle β with respect to the applied field $\mathbf{H}_0$ (see Figure 3.7), the direction cosines in Eq. 3.2 vary during each rotation period. For rapid sample rotation the time average of Eq. 3.2 becomes

$$\sigma = \tfrac{1}{2}\sin^2\beta\left(\sigma_{11} + \sigma_{22} + \sigma_{33}\right) + \tfrac{1}{2}\left(3\cos^2\beta - 1\right)$$

$$\times(\text{functions of direction cosines}). \qquad (3.4)$$

When the angle β between the axis of sample rotation and the applied magnetic field is 54.7° (the magic angle), $\sin^2\beta = \tfrac{2}{3}$, $3\cos^2\beta - 1 = 0$, and thus $\sigma = \sigma_i$, the isotropic chemical shift. Magic-angle spinning (MAS) reduces the anisotropic chemical-shift powder pattern (see Figure 3.6) to the isotropic average σ_i. In Figure 3.5(c) the broad overlapping carbonyl and aromatic carbon chemical-shift anisotropies of Figure 3.5(b) are reduced to their isotropic averages, leading to a truly high-resolution spectrum.

3.4.3. Cross-Polarization

In addition to high-power ^{13}C–^{1}H spin–spin coupling and chemical-shift anisotropy, one other obstacle must be overcome before high-resolution solid-state ^{13}C NMR spectra can be practically obtained. As mentioned in our discussion of high-resolution solution ^{13}C NMR performed by pulsed FT techniques (Sections 2.3 and 3.3), the rate at which signal averaging can be repeated, or the pulse repetition rate, is dictated by the T_1-values of the ^{13}C nuclei. Because most solids exhibit little motion in the megahertz frequency range, which is required for coupling of the spins to their surrounding nuclei or to the lattice, ^{13}C T_1-values are long for solids. Rare nuclei, such as ^{13}C (1.1% natural abundance), require signal averaging, and the repetition rate of rf pulses becomes an important consideration in their observation by NMR.

How can we circumvent the long signal accumulation times required by the low repetition rate for ^{13}C nuclei with long T_1's in solid samples? The answer lies in the ability to transfer the polarization of abundant ^1H nuclear spins with short T_1's to the rare ^{13}C nuclei. The repetition rate for signal averaging is now determined by the short ^1H T_1's, because energy is being transferred from the protons to the carbons. This process of polarization transfer from abundant to rare spins is termed cross-polarization (CP) and was introduced by Pines et al. (1972a, b).

Although ^1H and ^{13}C nuclei have Larmor frequencies different by a factor of four, Hartmann and Hahn (1962) showed that energy may be transferred between them in the rotating reference frame when

$$\gamma_C H_{1C} = \gamma_H H_{1H}. \tag{3.5}$$

Equation 3.5 results in a match of the rotating-frame energies for ^1H and ^{13}C and is called the Hartmann–Hahn condition. The match is produced when the applied carbon rf field (H_{1C}) is four times the strength of the applied proton rf field (H_{1H}), because $\gamma_H/\gamma_C = 4$. In Figure 3.8 the vector diagrams and pulse sequence are presented (Jelinski, 1982) for this double rotating-frame experiment which results in CP.

The proton and carbon spin systems are equilibrated in the magnetic field in Figure 3.8(a). An rf pulse H_{1H} at the proton Larmor frequency is applied along the x'-axis in (b) for a duration sufficient to tip the proton magnetization 90° along the y'-axis. The proton spins are forced to precess about the y'-axis of their rotating frame with frequency $\omega_H = \gamma_H H_{1H}$ for the duration of the strong H_{1H} pulse (c), a process called *spin locking*. During the time of spin locking of the proton nuclei, the carbon rf field H_{1C} is applied, thereby establishing contact between the two types of nuclei. ^{13}C magnetization grows in the direction of the spin-lock field (y'-axis) as the carbons precess about this axis with frequency $\omega_C = \gamma_C H_{1C}$.

Polarization is transferred between the proton and carbon nuclei as they both precess about the y'-axis by adjusting the power levels of the applied fields H_{1H} and H_{1C} until the Hartmann–Hahn condition is matched ($\gamma_H H_{1H} = \gamma_C H_{1C}$). The transfer of polarization is made possible because the z'-components of both ^1H and ^{13}C magnetizations have the same time dependence [Figure 3.8(c)], resulting in mutual spin flips. CP can be thought of as a "flow" of polarization from the abundant ^1H spins to the rare ^{13}C spins.

Since the ^{13}C nuclei obtain their polarization from the ^1H spins, it is the proton T_1 which determines the repetition rate of the CP experiment. This circumvents the problem of the long ^{13}C T_1's normally found in solids. In addition the ^{13}C signal shows an enhancement in its intensity, which can be as large as $\gamma_H/\gamma_C = 4$. The CP experiment results in both a time saving and an improvement in the signal-to-noise ratio in the ^{13}C NMR spectroscopy of solid samples.

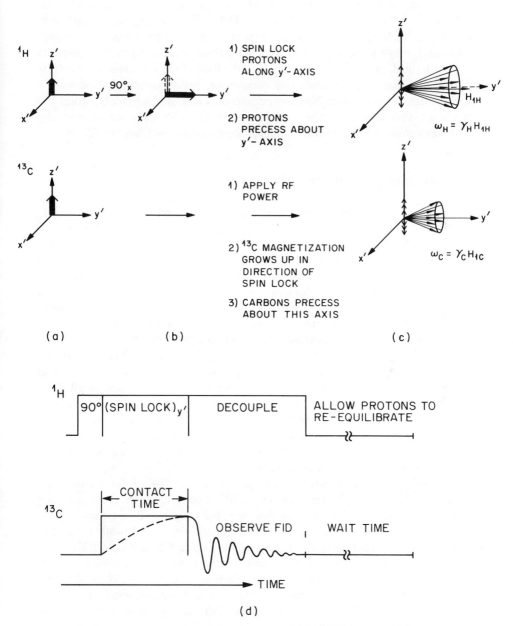

Figure 3.8 ■ (a), (b), and (c) are vector diagrams for a ^{1}H and ^{13}C double rotating-frame CP experiment. (d) gives the CP pulse sequence. [Reprinted with permission from Jelinski (1982).]

The first truly high-resolution NMR spectra of solid polymers were reported by Schaefer and Stejskal (1976). They combined the three previously
developed techniques of high-power proton decoupling, cross-polarization,
and magic-angle sample spinning to achieve these spectra. Since their pioneer-
ing work much progress has been made in the field of high-resolution solid-state
NMR, including the availability of commercial spectrometers that perform a
wide variety of solid-state NMR experiments. These developments permit the
study of the structures and conformations of solid polymer samples by
high-resolution NMR. Several examples are presented in Chapter 11.

3.5. Two-Dimensional NMR

If, instead of transforming (FT) the free induction decay (FID) immediately
after the 90° rf pulse in the usual way (see Section 2.3), we allow a time
interval for the nuclear spins to precess in the transverse plane and for the
evolution of interactions between them, then it is possible to obtain important
information concerning the nuclear spin system. We may divide such an
experiment into three time domains as indicated in Figure 3.9. The nuclear
spins are permitted to equilibrate with their surroundings via spin–lattice
relaxation during the preparation period. Following the $90°_x$ rf pulse, the x, y,
and z components of the nuclear spins evolve under all the forces acting upon
them, including their direct through-space dipole–dipole and through-bonds
scalar (J) couplings. This time domain, t_1, is termed the evolution period and
provides, along with the acquisition or detection time t_2 common to all pulse
experiments, the two-dimensional (2D) character of this experiment. Systematic incrementation of the evolution time t_1 (see Figure 3.9) provides the
second time dependence. After each t_1 period a second $90°_x$ rf pulse is
applied, and the exchange of nuclear spin magnetization may occur. The FID
is acquired during t_2 and transformed.

The pulse sequence presented in Figure 3.9 is appropriate for the observation of a chemical shift correlated or COSY spectrum, where the correlating
influence between nuclear spins is their scalar J-coupling. In a typical experiment we might utilize 1 K, or 1024, t_1-increments, with $t_1 = 0.5$–500 msec. The
FID following each t_1 is different because the interacting spins modulate each
other's response. Each FID detected in t_2 is transformed, producing a series of
1024 matrix rows, one for each t_1-value. Each row may consist of 1024 points
(square data matrix) representing the frequency-domain spectrum for a particular value of t_1, while the columns provide information about how the FIDs
were modulated as a function of t_1.

1024 new FIDs are constructed by looking down the columns of the data
matrix in an operation called the "transpose" in Figure 3.9. (Note that at this
stage the spectrum is represented for simplicity as a single resonance.) A
second Fourier transformation is performed on the newly transposed FIDs,

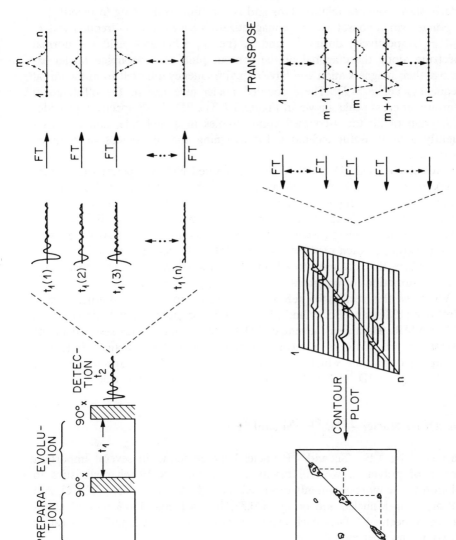

Figure 3.9 ■ Schematic representation of a two-dimensional (2D) correlated (COSY) experiment and spectrum, after Jelinski (1984). The correlating influence of the J-coupling between nuclei of different chemical shifts is shown. [Reprinted with permission from Jelinski (1984).]

leading to a 2D data matrix which is actually a surface in three-dimensional space. The surface may be represented as either a stacked plot or a contour plot. The contour plot is usually preferred, because the stacked plot does not clearly show complex relationships and is very time-consuming to record.

Nuclei which do not exchange magnetization have the same frequencies, F_1 and F_2 respectively, during t_1 and t_2 (i.e. $F_1 = F_2$) and yield the normal spectrum along the diagonal of the contour plot. Scalar-coupled nuclei exchange their magnetization and have a final frequency different from the initial frequency, i.e. $F_1 \neq F_2$. These coupled nuclei give rise to the off-diagonal contours or cross peaks shown in Figure 3.9. The 2D COSY spectrum provides a diagram of all the J-coupled connectivities in a molecule, and is consequently a very useful technique for assigning the resonances of complex molecules.

A closely related 2D NMR technique, termed NOESY, permits the establishment of through-space connectivities. This technique relies on the through-space coupling of nuclear spins and uses a 2D version of the nuclear Overhauser effect (NOE) (see Section 3.3) to map, in effect, all intra- and intermolecular distances (usually interproton) less than about 4 Å. We present an application of each of these powerful 2D NMR techniques in Chapters 6 and 8, where they are used to advantage in the study of the microstructures and conformations of polymers.

We close this section by referring the reader to Bax and Lerner (1986), Wüthrich (1986), and Bovey and Mirau (1988) for more complete descriptions of 2D NMR. The recent advent of 2D NMR techniques has resulted in the rebirth of ^{1}H NMR as a means to study molecular structure. Extensive J-coupling of protons, which unduly complicate 1D ^{1}H NMR spectra, are used to advantage in 2D ^{1}H NMR to map the connectivity of molecules.

3.6. Other Nuclei — ^{15}N, ^{19}F, ^{29}Si, and ^{31}P

Because ^{15}N, ^{19}F, ^{29}Si, and ^{31}P nuclei may be found in several important classes of polymers, we will briefly outline their NMR characteristics in relation to the more commonly observed ^{1}H and ^{13}C nuclei. ^{15}N, ^{19}F, ^{29}Si, and ^{31}P are spin-$\frac{1}{2}$ nuclei, and occur in 0.37, 100, 4.7, and 100% natural abundance, respectively. They each exhibit a range of chemical shifts at least as broad as observed for ^{13}C nuclei.

The ^{19}F and ^{31}P nuclei, because of their high natural abundance and large magnetogyric ratios, are nearly as easy to observe as ^{1}H. Like ^{1}H nuclei, ^{19}F and ^{31}P nuclei show extensive homonuclear scalar J-coupling and, in addition, ^{19}F–^{1}H and ^{31}P–^{1}H heteronuclear J-coupling (Emsley and Phillips, 1971; Crutchfield et al., 1967). It will be seen in subsequent chapters (6 and 8) that the ^{19}F chemical shifts of fluoropolymers are much more sensitive to their microstructures than are their ^{13}C chemical shifts. ^{19}F NMR is therefore the method of choice for studying fluoropolymers.

Though more abundant than ^{13}C (4.7 vs. 1.1%), the ^{29}Si nucleus has an even smaller nuclear dipole and magnetogyric ratio (γ) than ^{13}C (Marsmann, 1981). When this information is coupled with its 4.7% natural abundance, we would expect ^{29}Si resonances to be about twice as sensitive as ^{13}C resonances. However, because the magnetic moment and spin of the ^{29}Si nucleus are antiparallel, γ is negative. When broad-band proton decoupling is used to remove ^{29}Si–^{1}H scalar coupling, instead of an enhancement in the signal, as observed in ^{13}C NMR, the ^{29}Si signal may be reduced in intensity. In addition, the spin–lattice relaxation times for ^{29}Si nuclei in the dissolved state are typically rather long, much like those of ^{13}C nuclei. Nevertheless, pulsed FT NMR techniques have made the ^{29}Si nucleus, as well as ^{13}C, a valuable probe of silicon polymer microstructure.

Because its inherent sensitivity and natural abundance are substantially less than for ^{13}C, ^{15}N NMR spectroscopy is most often performed on enriched samples (Martin et al., 1982). However, pulsed FT NMR techniques have made the ^{15}N nucleus observable at natural abundance even in the solid state (see Chapter 11).

References

Alpert, N. L. (1947). *Phys. Rev.* **72**, 637.

Andrews, E. R., Bradbury, A., and Eades, R. G. (1959). *Nature (London)* **183**, 1802.

Bax, A. and Lerner, L. (1986). *Science* **232**, 960.

Bloch, F., Hansen, W. W., and Packard, M. E. (1946). *Phys. Rev.* **69**, 127.

Bovey, F. A., Tiers, G. V. D., and Filipovich, G. (1959). *J. Polymer Sci.* **38**, 73.

Bovey, F. A. and Mirau, P. A. (1988). *Accts. Chem. Res.* **21**, 37.

Crutchfield, M. M., Dungan, C. H., Letcher, J. H., Mark, V., and Van Wazer, J. R. (1967). *^{31}P Nuclear Magnetic Resonance*, Wiley-Interscience, New York.

Derome, A. E. (1987). *Modern NMR Techniques for Chemistry Research*, Pergamon, New York, Chapter 6.

Emsley, J. and Phillips, L. (1971). *Prog. in Nucl. Magn. Reson. Spect.* **7**, 1.

Farrar, T. C. and Becker, E. D. (1971). *Pulse and Fourier Transform NMR*, Academic Press, New York.

Ferguson, R. C. (1967a). *Polymer Preprints* **8** (2), 1026.

Ferguson, R. C. (1967b). *Trans. N.Y. Acad. Sci.* **29**, 495.

Flory, P. J. (1954). *Principles of Polymer Chemistry*, Cornell University Press, Ithaca, N.Y., Chapter XII.

Hartmann, S. R. and Hahn, E. L. (1962). *Phys. Rev.* **128**, 2042.

Heatley, F. and Zambelli, A. (1969). *Macromolecules* **2**, 618.

Jelinski, L. W. (1982). In *Chain Structure and Conformation of Macromolecules*, F. A. Bovey, Ed., Academic Press, New York, Chapter 8.

Jelinski, L. W. (1984). *Chem. Eng. News*, Nov. 5, p. 26.

Kuhlmann, K. F. and Grant, D. M. (1968). *J. Am. Chem. Soc.* **90**, 7355.

Marsmann, H. (1981). "^{29}Si-NMR Spectroscopic Results," in *NMR Basic Principles and Progress*, Vol. 17, P. Diehl, E. Fluck, and K. Kosfeld, Eds., Springer-Verlag, New York, p. 65.

Martin, G. J., Martin, M. L., and Gouesnard, J.-P. (1982). "^{15}N-NMR Spectroscopy," in *NMR Basic Principles and Progress*, Vol. 18, P. Diehl, E. Fluck, and K. Kosfeld, Eds., Springer-Verlag, New York, p. 1.

Morawetz, H. (1975). *Macromolecules in Solution*, Second Ed., Wiley-Interscience, New York, Chapter II.

Odajima, A. (1959). *J. Phys. Soc. Jap.* **14**, 777.

Pines, A., Gibby, M. G., and Waugh, J. S. (1972a). *J. Chem. Phys.* **56**, 1776.

Pines, A., Gibby, M. G., and Waugh, J. S. (1972b). *Chem. Phys. Lett.* **15**, 373.

Purcell, E. M., Torrey, H. C., and Pound, R. V. (1946). *Phys. Rev.* **69**, 37.

Saunders, M. and Wishnia, A. (1958). *Ann. N.Y. Acad. Sci.* **70**, 870.

Saunders, M., Wishnia, A., and Kirkwood, J. G. (1957). *J. Am. Chem. Soc.* **79**, 3289.

Schaefer, J. and Stejskal, E. O. (1976). *J. Am. Chem. Soc.* **98**, 1031.

Schilling, F. C., Bovey, F. A., Bruch, M. D., and Kozlowski, S. A. (1985). *Macromolecules* **18**, 1418.

Stothers, J. B. (1972). *Carbon-13 NMR Spectroscopy*, Academic Press, New York, Chapter 2.

Tonelli, A. E. and Schilling, F. C. (1981). *Accts. Chem. Res.* **14**, 233.

Wüthrich, K. (1986). *NMR of Proteins and Nucleic Acids*, Wiley, New York.

^{13}C NMR of Polymers

4.1. Introduction

Of the two nuclei ^{1}H and ^{13}C, which possess spin and are common to synthetic polymers, ^{13}C is by far the more sensitive spin probe for polymer NMR studies. ^{13}C NMR spectra suffer neither from a narrow dispersion of chemical shifts nor from extensive homonuclear, scalar spin–spin coupling, which both complicate the analysis of ^{1}H NMR spectra. In the previous chapter (see Sections 3.2 and 3.3) the superior resolution and sensitivity to microstructure exhibited by ^{13}C NMR polymer spectra were demonstrated by comparison of ^{13}C and ^{1}H NMR spectra recorded for polypropylene (PP) samples with different tacticities. There it was observed that the ^{13}C chemical shifts are spread over a 30-ppm range, while all the ^{1}H chemical shifts differ by less than 1 ppm.

It is this sensitivity of ^{13}C chemical shifts, $\delta\,^{13}$C, to the microstructures of molecules which makes ^{13}C NMR spectroscopy so useful as a structural probe. In Figure 3.4 we noted that the methyl carbon resonances observed in the 25-MHz ^{13}C NMR spectrum of atactic PP were sensitive to pentad stereosequences, and in the next chapter we will see that at higher field strength (90.5 MHz) the methyl carbon resonances show sensitivity to heptad stereosequences. The ^{13}C NMR spectra of PP are sensitive to stereosequences extending over 4 (pentads) and 6 (heptads) backbone bones. This long-range sensitivity to microstructural detail makes ^{13}C NMR a valuable tool in the determination of polymer structure.

To realize the full potential of ^{13}C NMR in microstructural studies the connections between microstructural features and the corresponding chemical shifts must be established. It is the purpose of the present chapter to discuss and establish these connections, so that we might be able to predict the ^{13}C NMR chemical shifts expected for each type of carbon atom in all possible structural environments. This permits the straightforward analysis of polymer

^{13}C NMR spectra and leads directly to a complete microstructural characterization without recourse to the synthesis and spectroscopic analysis of model compounds and polymers of known microstructure.

4.2. ^{13}C Chemical Shifts and Their Dependence on Microstructure

4.2.1. ^{3}C Nuclear Shielding

We have seen (Section 2.2.3) that the magnetic field H_i required to obtain the resonance condition for nucleus i at a particular irradiating rf frequency (H_1) is not equal to the applied field H_0, but instead is given by

$$H_i = H_0(1 - \sigma_i) \tag{4.1}$$

where the screening constant, σ_i, depends on the chemical structural environment of nucleus i. It is the cloud of electrons moving about the nucleus which shields it from the applied field H_0 by producing small local magnetic fields. Any structural feature that alters the electronic environment of a nucleus will affect its screening constant σ and lead to an alteration in its chemical shift δ at resonance.

To predict the chemical shift of a ^{13}C nucleus in a particular molecular environment the electronic wave function of the molecular system in the presence of the applied magnetic field must be known. For this reason it has been extremely difficult to make a priori predictions of ^{13}C NMR chemical shifts [see for example Ditchfield (1976) and Schastnev and Cheremisin, (1982)]. As an example, if we wish to calculate the relative chemical shifts of the ^{13}C nuclei in methane and methylfluoride, we must be able to determine the electronic wave functions of both molecules in the presence of H_0.

To date it has not been possible to make accurate predictions of ^{13}C NMR chemical shifts even when applying the most sophisticated ab initio quantum-mechanical methods. Instead the effects of substituents and local conformation have been used to correlate the ^{13}C chemical shifts and the microstructures of molecules, including polymers (Duddeck, 1986).

4.2.2. Substituent Effects on ^{13}C Chemical Shifts

^{13}C NMR studies of paraffinic hydrocarbons (Spiesecke and Schneider, 1961; Grant and Paul, 1964; Lindeman and Adams, 1971; Dorman et al., 1974) have led to substituent rules useful in the prediction of their ^{13}C chemical shifts. ^{13}C chemical shifts are ordered in terms of the effects produced by substituents attached to the observed carbon at the α, β, and γ positions.

In Table 4.1 we see the effect on the $\delta\,^{13}$C of the observed carbon (C°) when adding α-carbons to it. We observe a regular deshielding of about 9 ppm for each α-carbon added, except for neopentane, where the effect is reduced, presumably due to steric crowding. The effect of adding carbons β to the one

Table 4.1 ■ α-Substituent Effect on δ^{13}C (Bovey, 1974)

		δ^{13}C from TMS, ppm	α-effect, ppm
(a)	$^{o}CH_3$——H	-2.1	—
(b)	$^{o}CH_3$——$^{\alpha}CH_3$	5.9	8.0
(c)	$^{o}CH_2$ ($^{\alpha}CH_3$, $^{\alpha}CH_3$)	16.1	10.2
(d)	^{o}CH ($^{\alpha}CH_3$, $^{\alpha}CH_3$, $^{\alpha}CH_3$)	25.2	9.1
(e)	^{o}C ($^{\alpha}CH_3$, $^{\alpha}CH_3$, $^{\alpha}CH_3$, $^{\alpha}CH_3$)	27.9	2.7

observed, C°, also results in a deshielding of about 9 ppm, as is apparent from Table 4.2. On the other hand, a shielding of the observed carbon nucleus occurs when carbon substituents are added in the γ-position, and the magnitude of the γ-effect is reduced to about -2 ppm (see Table 4.3).

Using these substituent effects, i.e. $\alpha = \beta = +9$ ppm and $\gamma = -2$ ppm, it is possible to predict the δ^{13}C of a wide variety of paraffinic hydrocarbons, including highly branched compounds [see for example Lindeman and Adams, (1971)]. The assignment of resonances in the ^{13}C NMR spectra of paraffinic molecules to the corresponding carbon nuclei is materially assisted by these substituent effects.

As an example in polymer ^{13}C NMR spectra, let us suppose that polypropylene (PP) possesses occasional head-to-head (H–H) and tail-to-tail (T–T) units in addition to the predominant head-to-tail (H–T) enchainment of

Table 4.2 ■ β-Substituent Effect on $\delta^{13}C$ (Bovey, 1974)

		$\delta^{13}C$ from TMS, ppm	β-effect, ppm
(a)	$^{o}CH_3\!\!-\!\!\mid\!\!-\!^{\alpha}CH_3$	5.9	—
(b)	$^{o}CH_3\!\!-\!\!\mid\!\!-\!^{\alpha}CH_2\!\!-\!\!-\!^{\beta}CH_3$	15.6	9.7
(c)	$^{o}CH_3\!\!-\!\!\mid\!\!-\!^{\alpha}CH\big\langle\genfrac{}{}{0pt}{}{^{\beta}CH_3}{^{\beta}CH_3}$	24.3	8.7
(d)	$^{o}CH_3\!\!-\!\!\mid\!\!-\!^{\alpha}C\!\!-\!^{\beta}CH_3$ with $^{\beta}CH_3$, $^{\beta}CH_3$	31.5	7.2

monomer units. We can see from Figure 4.1 that in the H–H units methine carbons have one more β-substituent and two fewer γ-substituents than the H–T methines. We would therefore expect the H–H methine carbons to resonate $1\beta - 2\gamma = 1(9) - 2(-2) = 13$ ppm downfield from the H–T methine carbons. T–T methylene carbons have one less β-substituent and two more γ-substituents than the H–T methylene carbons, and the T–T methylene carbons would be expected to resonate $-\beta + 2\gamma = -1(9) + 2(-2) = -13$ ppm upfield from the H–T methylene carbons. H–H methyls have a single additional γ-substituent compared to H–T methyls, and they should come upfield $1\gamma = -2$ ppm from the H–T methyl resonances.

These expectations are borne out in the ^{13}C NMR spectra of H–T and H–H : T–T PPs (Schilling and Tonelli, 1980; Möller et al., 1981). $\delta^{13}C$'s observed in these spectra are summarized in Table 4.4 and are consistent with the $\delta^{13}C$'s predicted by the α-, β-, and γ-substituent effects.

4.2.3. The γ-Substituent Effect in ^{13}C NMR

We have just discussed the observation that γ-substituents in paraffinic hydrocarbons shield carbon nuclei relative to unsubstituted carbons. Because the observed carbon (C^{o}) and its γ-substituent (C^{γ}) are separated by three intervening bonds, their mutual distance and orientation are variable, depending on the conformation of the central bond. This is illustrated in the Newman projections presented in Figure 4.2. Note the distance between C^{o} and C^{γ}

Table 4.3 ■ γ-Substituent Effect on δ^{13}C (Bovey, 1974)

	$\delta^{13}C$ from TMS, ppm	γ-effect, ppm
(a) $^{o}CH_3\!-\!\!-^{\alpha}CH_2\!-\!\!-^{\beta}CH_3$	15.6	—
(b) $^{o}CH_3\!-\!\!-^{\alpha}CH_2\!-\!\!-^{\beta}CH_2\!-\!\!-^{\gamma}CH_3$	13.2	−2.4
(c) $^{o}CH_3\!-\!\!-^{\alpha}CH_2\!-\!\!-^{\beta}CH\!\!<^{\gamma}CH_3_{\gamma}CH_3$	11.5	−1.7
(d) $^{o}CH_3\!-\!\!-^{\alpha}CH_2\!-\!\!-^{\beta}C(\!^{\gamma}CH_3)(\!^{\gamma}CH_3)\!-\!\!^{\gamma}CH_3$	8.7	−2.8
(e) $^{\alpha}CH_3\!-\!\!-^{o}CH_2\!-\!\!-^{\alpha}CH_2\!-\!\!-^{\beta}CH_3$	25.0	
(f) $^{\alpha}CH_3\!-\!\!-^{o}CH_2\!-\!\!-^{\alpha}CH_2\!-\!\!-^{\beta}CH_2\!-\!\!-^{\gamma}CH_3$	22.6	−2.4
(g) $^{\alpha}CH_3\!-\!\!-^{o}CH_2\!-\!\!-^{\alpha}CH_2\!-\!\!-^{\beta}CH\!\!<^{\gamma}CH_3_{\gamma}CH_3$	20.7	−1.9
(h) $^{\alpha}CH_3\!-\!\!-^{o}CH_2\!-\!\!-^{\alpha}CH_2\!-\!\!-^{\beta}C(\!^{\gamma}CH_3)(\!^{\gamma}CH_3)\!-\!\!^{\gamma}CH_3$	18.8	−1.9

$(d_{o-\gamma})$ is reduced from 4 to 3 Å on changing their arrangement from *trans* to *gauche*.

Grant and Cheney (1967) first suggested the conformational origin of the γ-substituent effects on δ^{13}C. In their model it is the polarization of the C^{o}–H and C$^{\gamma}$–H bonds, resulting from their compression caused by proton–proton (o–γ) repulsion, that leads to a shielding of both carbon nuclei. More recently

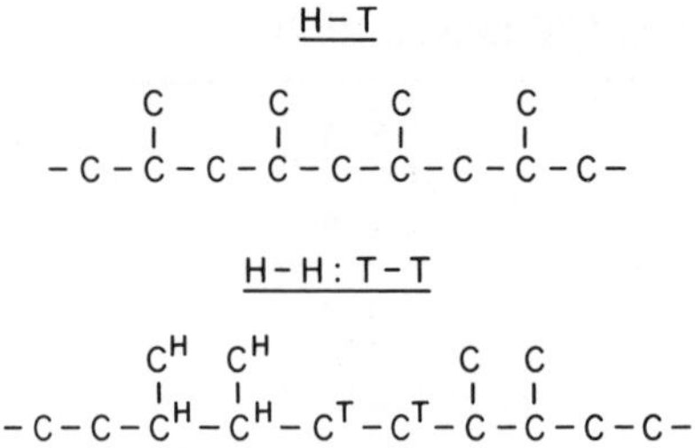

H–T

$$-C-\overset{\overset{\displaystyle C}{|}}{C}-C-\overset{\overset{\displaystyle C}{|}}{C}-C-\overset{\overset{\displaystyle C}{|}}{C}-C-\overset{\overset{\displaystyle C}{|}}{C}-C-$$

H–H : T–T

$$-C-C-\overset{\overset{\displaystyle C^H}{|}}{C^H}-\overset{\overset{\displaystyle C^H}{|}}{C^H}-C^T-C^T-\overset{\overset{\displaystyle C}{|}}{C}-\overset{\overset{\displaystyle C}{|}}{C}-C-C-$$

Figure 4.1 ■ Possible regiosequence of monomer units in PP.

Table 4.4 ■ $\delta^{13}C$'s Observed in the Spectra of H–T and H–H : T–T PPs

Carbon	$\delta^{13}C$ vs. TMS,[a] ppm		
	H–T[b]	H–H[c]	T–T[c]
CH	28.5	37.0	—
CH_2	46.0	—	31.3
CH_3	20.5	15.0	—

[a]All $\delta^{13}C$ values are averaged over the different stereosequences.
[b]Schilling and Tonelli (1980).
[c]Möller et al. (1981).

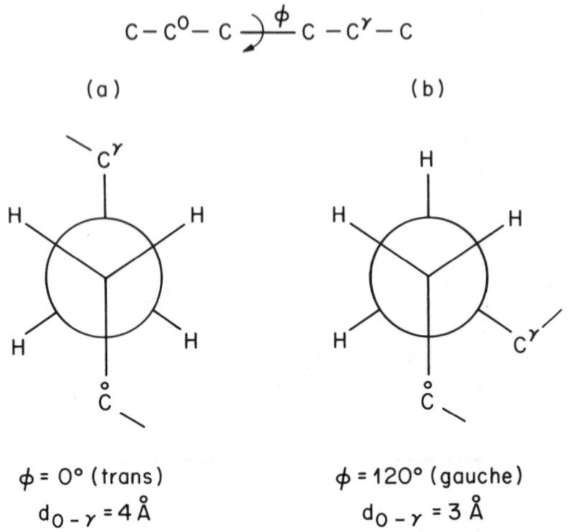

$$C-C^O-C \overset{\phi}{\rightarrow} C-C^\gamma-C$$

(a)   (b)

$\phi = 0°$ (trans)
$d_{O-\gamma} = 4 \, \text{Å}$

$\phi = 120°$ (gauche)
$d_{O-\gamma} = 3 \, \text{Å}$

Figure 4.2 ■ Newman projections of a *n*-alkane chain in the (a) *trans* ($\phi = 0°$) and (b) *gauche* ($\phi = 120°$) conformations.

Li and Chestnut (1985) have presented evidence that correlate shielding γ-effects with attractive van der Waals forces and not repulsive steric interactions, though their results still suggest that the *gauche* arrangement of the observed carbon and its γ-substituent is required for shielding. Seidman and Maciel (1977), using semiempirical and ab initio quantum-mechanical calculations, concluded that the γ-substituent effect is conformational in origin, but cannot be attributed solely to the proximity of the interacting C° and C^γ groups. It seems apparent then that the γ-substituent effect on $\delta^{13}C$ has a conformational origin and is, as we will shortly demonstrate, potentially useful in characterizing both the conformations and microstructures of polymers.

For a carbon nucleus to be shielded by a γ-substituent we have suggested that they must be in a *gauche* arrangement (see Figure 4.2). This suggestion is supported by comparing the $\gamma^{13}C$'s observed for the methyl carbons in *n*-alkanes. The methyl carbons in *n*-butane and higher *n*-alkanes have a single γ-substituent, while the methyl carbons in *n*-propane have no γ-substituents but the same number and kinds of α- and β-substituents as the higher *n*-alkanes. In their crystals the *n*-alkanes adopt the extended, all-*trans* conformation where both methyl carbons are *trans* to their γ-substituents. If the γ-substituents are *trans* to the methyl carbons in the higher solid *n*-alkanes, then we would expect δCH_3(solid C_2H_{2n+2}, $n \geq 4$) = δCH_3(liquid *n*-propane). VanderHart (1981) has observed the methyl carbons in the solid *n*-alkanes with $n = 19, 20, 23, 32$ to resonate between 15–16 ppm, while the methyl carbon in liquid *n*-propane resonates at 15.6 ppm (Stothers, 1972).

On the other hand, in the liquid state the methyl carbons in the higher *n*-alkanes ($n \geq 4$) resonate upfield at 13.2–14.1 ppm (Stothers, 1972). Of course in the liquid state the C–C bonds in *n*-alkanes possess a significant *gauche* content, and this results in the shielding of δCH_3 for *n*-butane and higher *n*-alkanes compared to that observed for the methyl carbons in *n*-propane or the higher solid *n*-alkanes in the all-*trans* conformation.

If we know how much *gauche* character, P_g, is possessed by the central bond between C° and X^γ

$$(C^\circ - C \overset{\phi}{\underset{}{\diagdown}} C - X^\gamma),$$

then we can estimate the shielding produced by X^γ, γ_{C-X}, when in a *gauche* arrangement with C°. This procedure is illustrated in Figure 4.3, where the *gauche* shielding effects of the γ-substituents C, OH, and Cl are derived. As an example, the shielding produced at the methyl carbon in *n*-butane by the other methyl group (its γ-substituent), i.e. $\Delta\delta CH_3 = \delta CH_3$(*n*-butane) $-$ δCH_3(*n*-propane) $= 13.2 - 15.6 = -2.4$ ppm, is divided by the *gauche* character of the intervening bond, $P_g = 0.46$: $\gamma_{C-C} = \Delta\delta CH_3/P_g = -2.4/0.46 = -5.2$ ppm. (In the next chapter we will describe the methodology used to calculate the bond conformational populations.)

When this procedure is applied to *n*-butane, 1-propanol, and 1-chloropropane, the following γ-*gauche* shielding effects are derived: $\gamma_{C-C} = -5.2$ ppm,

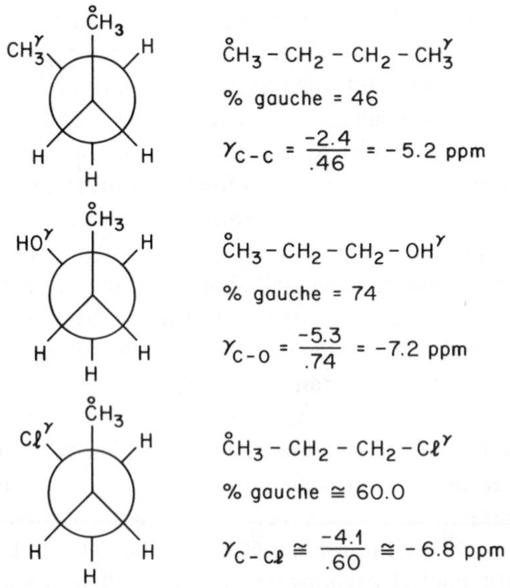

$$\overset{\circ}{C}H_3 - CH_2 - CH_2 - CH_3^\gamma$$

% gauche = 46

$$\gamma_{C-C} = \frac{-2.4}{.46} = -5.2 \text{ ppm}$$

$$\overset{\circ}{C}H_3 - CH_2 - CH_2 - OH^\gamma$$

% gauche = 74

$$\gamma_{C-O} = \frac{-5.3}{.74} = -7.2 \text{ ppm}$$

$$\overset{\circ}{C}H_3 - CH_2 - CH_2 - Cl^\gamma$$

% gauche $\cong$ 60.0

$$\gamma_{C-Cl} \cong \frac{-4.1}{.60} \cong -6.8 \text{ ppm}$$

Figure 4.3 ■ Derivation of the γ-*gauche* shielding produced by the γ-substituents C, OH, and Cl (see text).

$\gamma_{C-O} = -7.2$ ppm, and $\gamma_{C-Cl} = -6.8$ ppm. We now see that the shielding at a carbon nucleus produced by a γ-substituent in a gauche arrangement can be comparable in magnitude (-5 to -7 ppm) to the deshielding ($+9$ ppm) caused by the more proximal α and β substituents. More important, however, is the conformational dependence of the γ-substituent effect on ^{13}C NMR chemical shifts. Any variation in the microstructure of a molecule which effects its local conformation can be expected to be reflected in its δ^{13}C's via the γ-*gauche* effect.

4.2.4. γ-*gauche* Effects in ^{13}C NMR

Let us close our discussion of ^{13}C NMR chemical shifts by illustrating the conformational connection between observed δ^{13}C's and microstructure. This connection is provided by the conformationally sensitive γ-*gauche* effect. In Table 4.5 we present the nonequivalent δ^{13}C's observed for the isopropyl methyl carbons in several branched alkanes. Even though the isopropyl methyl carbons in each alkane have the same α, β, and γ substituents, we note in column 2 that the observed nonequivalence progressively decreases as the number of carbons separating the terminal isopropyl group from the asymmetric center is increased. This behavior can be understood if we focus on the source of the nonequivalent δ^{13}C's observed for the isopropyl methyl carbons in 2,4-dimethylhexane (2,4-DMH).

Table 4.5 ▪ Nonequivalent ¹³C NMR Chemical Shifts for the Isopropyl
Methyl Carbons in Branched Alkanes

Alkane	$\Delta\delta$, ppm	
	Obsd.[a]	Calcd.
C C | | C—C—C—C—C—C	1.0 (1.9, 1.1, 0.9)[b]	1.6, 1.1, 0.9
C C | | C—C—C—C—C—C—C	0.2	0.2
C C | | C—C—C—C—C—C—C—C	0.1	0.04
C C | | C—C—C—C—C—C—C—C—C	0.0	0.0

[a] Observed between ambient temperature and 48°C. (Kroschwitz et al., 1969;
Lindeman and Adams, 1971; Carman et al., 1973).
[b] Observed at -120, 25, and 90°C (Tonelli et al., 1984).

In Figure 4.4 we have illustrated the possible conformations about the C_2–C_3 backbone bond in 2,4-DMH, since these determine whether or not the isopropyl methyl carbons C^{sc}, C^{bb} are γ-gauche to the asymmetric carbon C_4. From the probabilities of finding the bond C_2–C_3 in the trans (t), gauche$^+$ (g^+), and gauche$^-$ (g^-) rotational states (P_t, P_{g^+}, P_{g^-}), we obtain $P_t + P_{g^+}$ and $P_{g^+} + P_{g^-}$ as the probabilities for gauche arrangements between C^{sc} and C^{bb}, respectively, and their γ-substituent C_4. Bond rotation probabilities are obtained, as described in Chapter 5 (Tonelli et al., 1984), from the conformational model developed by Mark (1972) for ethylene–propylene copolymers: $P_t = 0.38$, $P_{g^+} = 0.01$, and $P_{g^-} = 0.61$. Thus, C_4 is γ-gauche to C^{sc} with probability 0.39 and to C^{bb} with probability 0.62. We expect the nonequivalence between C^{sc} and C^{bb} to be $\Delta\delta\,^{13}C = (0.39 - 0.62) \times \gamma_{C-C} = -0.23(-5 \text{ ppm})$ $= 1.1$ ppm, where we have adopted the value $\gamma_{C-C} = -5$ ppm derived from n-butane.

The observed nonequivalence (1.0–1.1 ppm) is in close agreement with the value expected from the γ-gauche conformational calculation. The temperature dependence of the observed magnetic nonequivalence is also successfully reproduced by the γ-gauche-effect calculations, leaving little doubt that its origin is the conformationally sensitive γ-gauche effect.

From the Newman projections in Figure 4.4(b) it might be expected that the t and g^- conformations would be equally populated. However, it is well known (Flory, 1969) that rotational-state probabilities for the bonds in linear chain molecules depend on the conformations, or rotational states, of neigh-

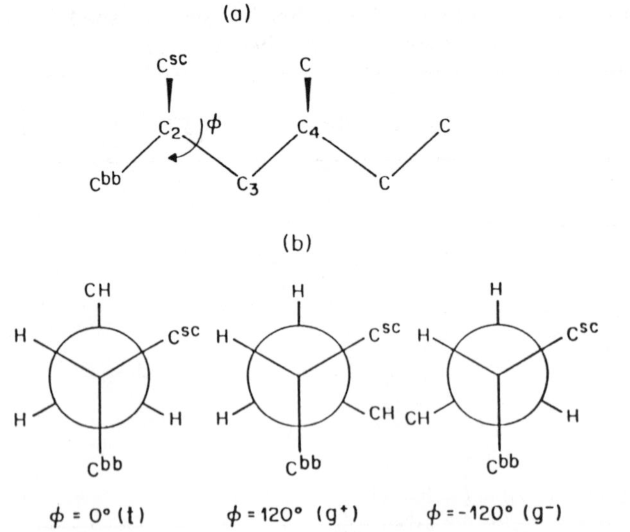

(a)

(b)

$\phi = 0°$ (t) $\phi = 120°$ (g⁺) $\phi = -120°$ (g⁻)

Figure 4.4 ▪ (a) 2,4-DMH in the all-*trans* conformation. (b) Newman projections illustrating rotational states about the C_2–C_3 backbone bond of 2,4-DMH. [Reprinted with permission from Tonelli et al. (1984).]

boring bonds. The asymmetric center at C_4 produces intramolecular interactions which depend simultaneously on ϕ and neighboring bond rotations [see Figure 4.4(a)] which render $P_t \neq P_{g^-}$. The values of P_t and P_{g^-} approach each other as the asymmetric center is further removed from the terminal isopropyl group, leading to a reduction in the expected nonequivalence of the isopropyl methyl carbons. This expectation is borne out in Table 4.5, where it is both observed and predicted that the magnetic nonequivalence of isopropyl methyl carbons vanishes once they are separated by more than four carbons from the asymmetric center.

It is apparent from this example that the microstructural sensitivity of ^{13}C NMR chemical shifts can have a conformational origin. $\delta\,^{13}$C depends on the local magnetic field, which is influenced by the local conformation in the vicinity of the resonating carbon nucleus. The local conformation is determined by the neighboring microstructure. Hence, the microstructural sensitivity of ^{13}C NMR has its basis in the dependence of the local conformation on microstructure.

Remembering back to the last chapter, where in Figure 3.4 we presented the 25-MHz ^{13}C NMR spectrum of atactic PP, we noted that the methyl-carbon region of the spectrum displayed sensitivity to pentad stereosequences. We are now in a position to understand this long-range sensitivity to microstructure and realize that it is caused by the dependence of the local conformation about a given methyl carbon to the neighboring stereosequences. The range of the microstructural dependence of $\delta\,^{13}$C's in PP (pentads) is comparable to that we observed and discussed in the small-molecule examples of the branched alkanes.

As we have seen, the γ-*gauche* effect can be utilized to make the connection between microstructure and ^{13}C NMR. However, sufficient conformational information, specifically how the local conformation depends on neighboring microstructure, is required to complete this connection. In the next chapter we outline the development of conformational models for polymers and describe in detail how we calculate ^{13}C NMR chemical shifts using the γ-*gauche* method.

References

Bovey, F. A. (1974). In *Proceedings of the International Symposium on Macromolecules, Rio de Janerio, July 26–31, 1974*, E. B. Mano, Ed., Elsevier, New York, p. 169.

Carman, C. J., Tarpley, A. R., Jr., and Goldstein, J. H. (1973). *Macromolecules* **6**, 719.

Ditchfield, R. (1976). *Nucl. Magn. Reson.* **5**, 1.

Dorman, D. E., Carhart, R. E., and Roberts, J. D. (1974). Private communication cited in Bovey (1974).

Duddeck, H. (1986). In *Topics in Stereochemistry*, Vol. 16, E. L. Eliel, S. H. Wilen, and N. L. Allinger, Eds., Wiley-Interscience, New York, p. 219.

Flory, P. J. (1969). *Statistical Mechanics of Chain Molecules*, Wiley-Interscience, New York.

Grant, D. M. and Cheney, B. V. (1967). *J. Am. Chem. Soc.* **89**, 5315.

Grant, D. M. and Paul, E. G. (1964). *J. Am. Chem. Soc.* **86**, 2984.

Kroschwitz, J. I., Winokur, M., Reid, H. J., and Roberts, J. D. (1969). *J. Am. Chem. Soc.* **91**, 5927.

Li, S. and Chestnut, D. B. (1985). *Magn. Reson. Chem.* **23**, 625.

Lindeman, L. P. and Adams, J. Q. (1971). *Anal. Chem.* **43**, 1245.

Mark, J. E. (1972). *J. Chem. Phys.* **57**, 2541.

Möller, M., Ritter, W., and Cantow, H.-J. (1981). *Polym. Bull.* **4**, 609.

Schastnev, P. V. and Cheremisin, A. A. (1982). *J. Struct. Chem.* **23**, 440.

Schilling, F. C. and Tonelli, A. E. (1980). *Macromolecules* **13**, 270.

Seidman, K. and Maciel, G. E. (1977). *J. Am. Chem. Soc.* **99**, 659.

Spiesecke, H. and Schneider, W. G. (1961). *J. Chem. Phys.* **35**, 722.

Stothers, J. B. (1972). *Carbon-13 NMR Spectroscopy*, Academic Press, New York, Chap. 3.

Tonelli, A. E., Schilling, F. C., and Bovey, F. A. (1984). *J. Am. Chem. Soc.* **106**, 1157.

VanderHart, D. L. (1981). *J. Magn. Reson.* **44**, 117.

γ-*gauche*-Effect Method of Predicting ^{13}C NMR Chemical Shifts

5.1. Introduction

In the previous chapter we attempted to establish the connections between the chemical shifts observed in the ^{13}C NMR spectra of polymers and their microstructures. The local magnetic field experienced by a carbon nucleus is sensitive to the local conformation in its vicinity. We have also seen that the local conformation is influenced by the microstructural environment of the observed carbon nucleus. Hence, a polymer's microstructure determines its δ^{13}C's through the influence it exerts on the local conformation, which modifies the resultant local magnetic field H_i:

$$\text{microstructure} \rightarrow \text{conformation} \rightarrow H_i \rightarrow \delta^{13}\text{C}.$$

The shielding of ^{13}C nuclei by γ-substituents in a *gauche* arrangement (γ-*gauche* effect) was introduced in the last chapter to make the connection between microstructure and δ^{13}C. In the present chapter we will discuss polymer conformations and develop methods for determining their populations as a function of polymer microstructure. This is necessary for the calculation of δ^{13}C's via the γ-*gauche*-effect method, which relies on a knowledge of bond conformational probabilities to determine the extent of γ-*gauche* shielding experienced by a ^{13}C nucleus in a given microstructural environment:

$$\text{microstructure} \xrightarrow[\text{effect}]{\gamma\text{-}gauche} \delta^{13}\text{C}.$$

5.2. Polymer Conformations

5.2.1. Rotational Isomeric State Model of Polymers

To specify the spatial relationships between the segments of a polymer chain, or the polymer's conformation, we must know the bond lengths l, the bond valence angles $\pi - \theta$, and the rotational states ϕ, of each bond in the chain (see Figure 5.1). For an all-carbon-backbone polymer like the one depicted in Figure 5.1, all bond lengths $l = 1.54$ Å and all valence angles $\pi - \theta = 112°$ (Flory, 1969). Rotations ϕ about the C–C backbone bonds remain as the principal degree of conformational freedom in polymers.

Spectroscopic (Mizushima, 1954; Wilson, 1959, 1962; Herschback, 1963) and electron-diffraction (Bonham and Bartell, 1959; Bartell and Kohl, 1963; Kuchitsu, 1959) studies of low-molecular-weight compounds have clarified the nature of the potentials hindering rotations about chemical bonds. Rotation about the C–C bond in ethane (and the higher n-alkanes) is threefold, with energy minima corresponding to the staggering of the methyl hydrogen atoms. The rotation potential $E(\phi_2)$ about the central C–C bond in n-butane is shown in Figure 5.2. As for ethane, the rotation potential is threefold but no longer symmetrical, owing to the nonbonded interactions of the terminal methyl groups in the gauche rotational states at $\phi_2 = \pm 120°$. (Rotation angles are taken as $0°$ in the planar zigzag or *trans* conformation and assume positive values for right-handed rotations.) The barrier separating the *gauche* ($\phi_2 = \pm 120°$) and *trans* ($\phi_2 = 0°$) rotational states in n-butane is about 3.5 kcal/mol (14.6 kJ/mol) (Herschback, 1963), which considerably exceeds RT at ordinary temperatures, where R is the gas constant. At equilibrium the distribution of rotation angles is clearly nonuniform over the 2π range, with $\phi_2 = 0°$, $\pm 120°$ being heavily preferred over $\phi_2 = 180°$, $\pm 60°$.

A realistic representation of the internal rotation state of n-butane is afforded by considering each molecule to be confined to small oscillations ($\pm 20°$) about each of its three energy minima $\phi_2 = 0°$, $\pm 120°$. [The barrier

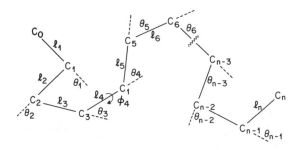

Figure 5.1 ■ Schematic representation of a carbon-backbone polymer chain. [Reprinted with permission from Tonelli (1986).]

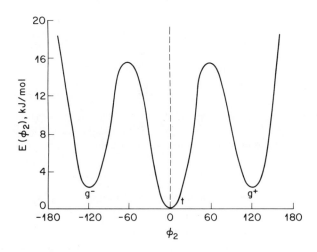

Figure 5.2 ▪ The potential energy $E(\phi_2)$ of rotation ϕ_2 about the central C–C bond in butane $(CH_3—CH_2\overset{\phi_2}{\mathcal{\mp}}CH_2—CH_3)$. Both methyl groups are fixed in the staggered positions. To convert J to cal, divide by 4.184. [Reprinted with permission from Tonelli (1986).]

separating the minima is not large enough to prevent rapid transitions (about 10^{10}/sec) between the minima, but does insure that at equilibrium a negligible population of n-butane molecules have $\phi_2 = 180°$, $\pm 60°$.]

The discrete nature of the rotational potential has been recognized (Volkenstein, 1963; Birshstein and Ptitsyn, 1966), and in the rotational isomeric state (RIS) approximation, each bond in the polymer backbone is assumed to occur in any one of a small number of discrete rotational states. These states are generally selected to coincide with the potential minima (see Figure 5.2). For polymers with bonds whose rotational barriers do not exceed RT, the RIS model may be used to approximate the nearly continuous rotation potential by a discrete sum of rotational states. [See Mansfield (1983) for a discussion of methods developed for treating polymers with low rotational energy barriers and for treating possible fluctuations in bond rotations.]

Not only is each bond in a polymer chain usually restricted to a few discrete rotational states, but the probability of occurrence of any rotational state about a given bond depends on the rotational states of its nearest-neighbor bonds (Flory, 1969; Volkenstein, 1963; Birshstein and Ptitsyn, 1966). This rotational interdependence is clearly illustrated by the conformations of n-pentane as presented in Figure 5.3. Severe repulsive steric interaction between the terminal methyls occurs in the g^+g^- (or g^-g^+) conformation of Figure 5.3(c), whereas in the g^+g^+ (or g^-g^-) conformation of Figure 5.3(b), the methyl groups are sufficiently separated (about 3.6 Å = 0.36 nm) to suggest an approximately neutral (neither attractive nor repulsive) interaction. The energy of the rotational state ϕ_2 of bond 2 evidently depends on the

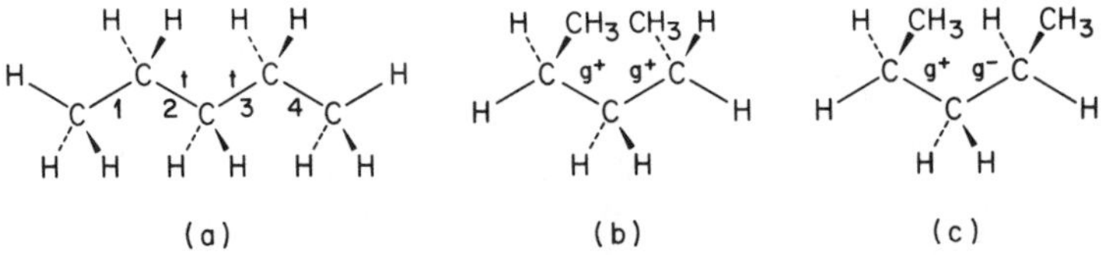

Figure 5.3 ■ *n*-Pentane conformations $(CH_3-CH_2 \overset{\phi_2}{\boldsymbol{\downarrow}} CH_2 \overset{\phi_3}{\boldsymbol{\downarrow}} CH_2-CH_3)$: (a) *trans,trans* $(\phi_2 = \phi_3 = 0°)$; (b) *gauche*$^+$, *gauche*$^+$ $(\phi_2 = \phi_3 = 120°)$; and (c) *gauche*$^+$, *gauche*$^-$ $(\phi_2 = 120°, \phi_3 = -120°)$. [Reprinted with permission from Tonelli (1986).]

rotational state ϕ_3 of bond 3 in *n*-pentane. When $\phi_3 = g^-$, ϕ_2 would prefer to be t or g^- rather than g^+, owing to the severe interaction between methyl groups in the $\phi_2, \phi_3 = g^+, g^-$ conformation.

In most polymer chains, as in *n*-pentane, the relative probabilities of the rotational states of a given bond depend on the rotational states of nearest-neighbor bonds. It is assumed that the locations, unlike the energies, of the rotational states about a given bond are unaffected by the rotational states of its neighbors. For most polymer chains, this assumption appears to be justified, and for those cases where both the locations and energies of rotational states are interdependent, additional rotational states can be incorporated (Flory, 1969). The nonbonded interactions between groups separated by two backbone bonds (see Figure 5.3) are the source of this nearest-neighbor interdependence of polymer-chain conformation. Thus, most polymer chains can be treated as one-dimensional (linear chains), statistical-mechanical systems composed of nearest-neighbor-dependent elements. Such systems are conveniently treated by the mathematical methods developed (Kramers and Wannier, 1941; Newell and Montroll, 1953) to handle the one-dimensional Ising model of ferromagnetism (Ising, 1925).

The RIS model successfully accounts for those equilibrium properties dependent upon the conformational and configurational characteristics of individual polymer chains (Flory, 1969; Tonelli, 1986). More importantly, its adoption eliminates the need to introduce artificial chain models which, though simpler, disregard chain geometry and chemical structure, the two distinguishing features of each polymer chain.

If all bond lengths and valence angles (see Figure 5.1) are assumed to be fixed, the conformation of a *n*-bond polymer chain can be specified by assigning a rotational state to each of the $n-2$ nonterminal bonds. Let ν be the number of rotational states (usually three) about each bond. Then there are ν^{n-2} possible conformations in all. For polyethylene, $+CH_2-CH_2+_{n/2}$, with a typical chain length $n = 10,000$, the total number of possible conformations is $3^{10,000} = 10^{4800}$, truly an astronomically large number. Because the barriers separating the rotational states in polyethylene are low (3.5 kcal/mol), there is rapid transition (10^{10} conformations/sec) between, and eventual sampling of, all possible conformations. The time required to sample all 10^{4800} possible conformations can be estimated as $(10^{4800}$ conformations$)/(10^{10}$ conformations/sec$) = 3 \times 10^{4782}$ years. Our universe is believed to be a mere 1–2×10^{10} years old. This leads to the interesting notion that if a 10,000-bond polyethylene chain were formed at the instant of the Big Bang genesis of our universe, then at present it would only have sampled a minute portion of all its possible conformations.

The energy $E\{\phi\}$ of any conformation can be expressed in terms of the pairwise, nearest-neighbor-dependent energies $E(\phi_{i-1}, \phi_i)$:

$$E\{\phi\} = \sum_{i=2}^{n-1} E_i(\phi_{i-1}, \phi_i) = \sum_{i=2}^{n-1} E_{\xi\eta;\, i}, \tag{5.1}$$

where ξ and η denote the rotational states of bonds $i-1$ and i, respectively. Statistical weights $\mu_{\xi\eta}$, or Boltzmann factors, corresponding to the energies $E_{\xi\eta}$ may be defined as

$$\mu_{\xi\eta;\, i} = \exp\left[-E_{\xi\eta;\, i}/RT\right] \tag{5.2}$$

and expressed in matrix form:

$$U_i = [\mu_{\xi\eta;\, i}] = \begin{bmatrix} \mu_{\alpha\alpha} & \mu_{\alpha\beta} & \cdots & \mu_{\alpha\nu} \\ \mu_{\beta\alpha} & \mu_{\beta\beta} & \cdots & \mu_{\beta\nu} \\ \vdots & \vdots & & \vdots \\ \mu_{\nu\alpha} & \mu_{\nu\beta} & \cdots & \mu_{\nu\nu} \end{bmatrix}. \tag{5.3}$$

The rows of the $\nu \times \nu$ matrix U_i are indexed with the states ξ of bond $i-1$, and its columns with the states η of bond i.

The statistical weight of a particular chain conformation is then simply

$$\Omega_{\{\phi\}} = \prod_{i=2}^{n-1} \mu_{\xi\eta;\, i}. \tag{5.4}$$

Summing Eq. 5.4 over all possible conformations leads formally to the configurational partition function

$$Z = \sum_{\{\phi\}} \Omega_{\{\phi\}} = \sum_{\{\phi\}} \prod_{i=2}^{n-1} \mu_{\xi\eta;\, i}. \tag{5.5}$$

Application (Flory, 1969) of matrix methods (Kramers and Wannier, 1941), which were previously used to treat the Ising ferromagnet (Ising, 1925), leads to

$$Z = J^* \left[\prod_{i=2}^{n-1} U_i\right] J, \tag{5.6}$$

where J^* and J are the $1 \times \nu$ and $\nu \times 1$ row and column vectors

$$J^* = \begin{bmatrix} 1 & 0 & \cdots & 0 \end{bmatrix}, \qquad J = \begin{bmatrix} 1 \\ 1 \\ 1 \\ \vdots \\ 1 \end{bmatrix}. \qquad (5.7)$$

Equations 5.6 and 5.7 permit evaluation of the configurational partition function for a polymer chain of any length provided its RIS model (i.e., the energies of the bond rotational states) is known.

How are the locations and energies of the rotational states determined for a given polymer? Ideally, independent spectroscopic, thermodynamic, and diffraction data could be used to establish the RIS models of polymers. However, with the exception of the n-alkanes, including polyethylene, data sufficient to establish the RIS models of polymers are not yet available. Instead, one of two other methods for establishing the conformational energies of a polymer must be used.

Semiempirical potential functions (Flory, 1969; Hendrickson, 1961; Abe et al., 1966; Borisova, 1964; Scott and Scheraga, 1966), which include terms to estimate the intrinsic torsional potential and nonbonded van der Waals and electrostatic interactions, can be used to calculate the conformational energies of polymers. As an example, the conformational energy map obtained by Abe et al. (1966) for n-pentane, which is presented in Figure 5.4, may be used to establish the RIS model of polyethylene (see Flory, 1969). Relative to the $\phi_2, \phi_3 = 0°, 0° = tt$ conformation, $E_{tg^\pm} = E_{g^\pm t} = 2.1$ kJ/mol, $E_{g^\pm g^\pm} = 4.2$ kJ/mol, and $E_{g^\pm g^\mp} = 13.4$ kJ/mol. When these energies are used to construct the Boltzmann factors or statistical weights $\mu_{\xi\eta; i}$ (see Eq. 5.2), which constitute the elements of the statistical weight matrix U_i (see Eq. 5.3), the resulting RIS model successfully describes the conformation-dependent properties of

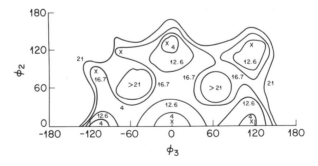

Figure 5.4 ■ Conformational energy map (Abe et al., 1966) for n-pentane,

$$CH_3 \overset{\phi_1}{\underset{\psi}{-}} CH_2 \overset{\phi_2}{\underset{\psi}{-}} CH_2 \overset{\phi_3}{\underset{\psi}{-}} CH_2 \overset{\phi_4}{\underset{\psi}{-}} CH_3 ,$$

with $\phi_1 = \phi_4 = 0°$. Minima are indicated by a X, and contours are shown in kJ/mol (cal = J/4.184). [Reprinted with permission from Tonelli (1986).]

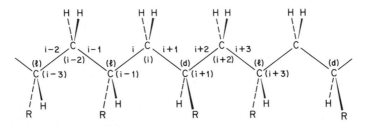

Figure 5.5 ■ Portion of an atactic vinyl chain. Asymmetric centers are given the designations *d* and *l* according to the arbitrary convention adopted in the text. Serial indexes referring to chain atoms are shown in parentheses. [Reprinted with permission from Flory (1969).]

polyethylene (Flory, 1969). Unfortunately, for many other polymers it appears at present that the inaccuracies inherent in semiempirical energy estimates are sufficient to preclude accurate evaluation of energy differences between rotational states, though their locations can usually be determined by such calculations.

The most widely employed method for determining the RIS model of a polymer chain utilizes rotational-state statistical weights as parameters to be determined by comparison of calculated and measured conformation-dependent properties, such as the mean-square end-to-end distance $\langle r^2 \rangle$. Let us illustrate this procedure for a vinyl polymer shown in Figure 5.5. Asymmetric centers with R groups above (below) the backbone plane are arbitrarily called *d* (*l*) centers. First-order interactions, which depend on a single rotation angle, are indicated in Figure 5.6. For rotations about $CH-CH_2$ bonds such as *i*, *t*, g^+, and g^-, states are assigned statistical weights $\mu_1 = \eta$, 1, and τ, where η is the ratio of statistical weights for $CH---R$ and $CH----CH_2$ interactions and τ accounts for simultaneous $CH---R$ and $CH---CH_2$ interactions. η,

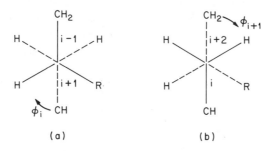

Figure 5.6 ■ Diagrams illustrating nonbonded interactions dependent on a single rotation angle: (a) ϕ_i, and (b) ϕ_{i+1}. The bonds in the background, i.e., those attached to the skeletal atom of higher index [$i + 1$ in (a) and $i + 2$ in (b)], are shown as dashed lines. Both asymmetric centers are represented in the *d*-configuration. [Reprinted with permission from Flory (1969).]

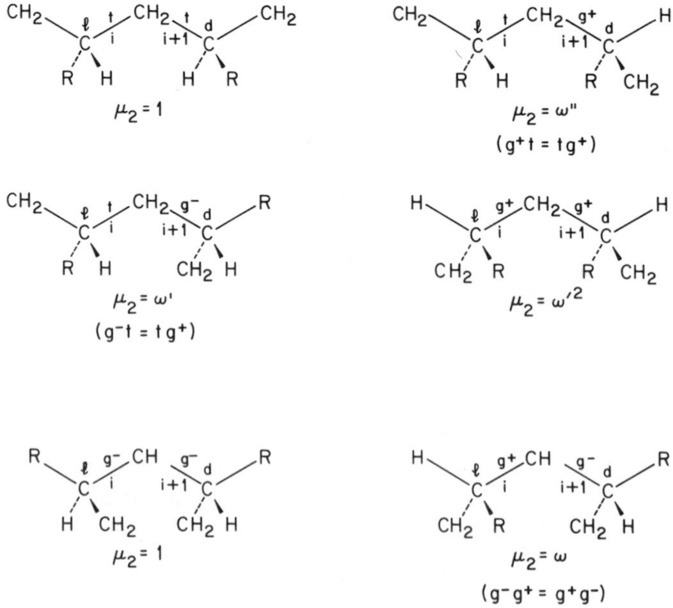

$\mu_2 = 1$

$\mu_2 = \omega''$

$(g^+ t = t g^+)$

$\mu_2 = \omega'$

$(g^- t = t g^+)$

$\mu_2 = \omega'^2$

$\mu_2 = 1$

$\mu_2 = \omega$

$(g^- g^+ = g^+ g^-)$

Figure 5.7 ▪ Second-order interactions dependent on bond rotations ϕ_i, ϕ_{i+1} (see Figure 5.5).

τ, and 1 are the first-order statistical weights describing the t, g^+, and g^- rotational states about CH_2—CH bonds such as $i+1$. For l-centers the first-order interactions and their corresponding statistical weights are reversed for the g^+ and g^- rotational states.

If we assign statistical weights ω, ω', and ω'' to the second-order interactions between CH_2---CH_2 or CH---CH, R---CH_2, and R---R, which depend on two consecutive bond rotations, then we may develop RIS statistical weight matrices for each of the bonds in the atactic vinyl polymer shown in Figure 5.5. In Figure 5.7 we have illustrated all second-order interactions μ_2 dependent on bond rotations ϕ_i, ϕ_{i+1}. Based on the first- and second-order interactions (μ_1, μ_2) and their corresponding statistical weights (see Figures 5.6 and 5.7), we may write for the bond pair ϕ_i, ϕ_{i+1} in Figure 5.5

$$
U_{i+1} = U_{ld} =
\begin{array}{c}
\phi_i/\phi_{i+1} \\
\\
t \\
g^+ \\
g^-
\end{array}
\begin{array}{c}
t \quad\quad g^+ \quad\quad g^- \\
\left[
\begin{array}{ccc}
\eta & \tau\omega'' & \omega' \\
\eta\omega'' & \tau\omega'^2 & \omega \\
\eta\omega' & \tau\omega & 1
\end{array}
\right]
\end{array}
\tag{5.8}
$$

where each element of U_{i+1} is a product of first-order interactions $\mu_1 = \eta, \tau, 1$ dependent on ϕ_{i+1} with second-order interactions $\mu_2 = 1$, ω, ω', ω'^2, and ω'' dependent on both ϕ_i and ϕ_{i+1}. For example, $U_{ld}(tg^+) = \mu_1(\phi_{i+1} = g^+) \times \mu_2(\phi_i = t, \phi_{i+1} = g^+) = \tau\omega''$.

For the other racemic diad (*dl*) and the two meso diads (*dd* and *ll*), the following statistical weight matrices may be derived (Flory, 1969; Bovey, 1982):

$$U_{dl} = \begin{bmatrix} \eta & \omega' & \tau\omega \\ \eta\omega' & 1 & \tau\omega \\ \eta\omega'' & \omega & \tau\omega'^2 \end{bmatrix}, \tag{5.9}$$

$$U_{ll} = \begin{bmatrix} \eta\omega'' & 1 & \tau\omega' \\ \eta\omega' & \omega' & \tau\omega\omega'' \\ \eta & \omega & \tau\omega' \end{bmatrix}, \tag{5.10}$$

$$U_{dd} = \begin{bmatrix} \eta\omega'' & \tau\omega' & 1 \\ \eta & \tau\omega' & \omega \\ \eta\omega' & \tau\omega\omega'' & \omega' \end{bmatrix}. \tag{5.11}$$

For bond pairs flanking the asymmetric centers two statistical weight matrices U_l and U_d can be similarly derived:

$$U_l = \begin{bmatrix} \eta & \tau & 1 \\ \eta & \tau & \omega \\ \eta & \tau\omega & 1 \end{bmatrix}, \tag{5.12}$$

$$U_d = \begin{bmatrix} \eta & 1 & \tau \\ \eta & 1 & \tau\omega \\ \eta & \omega & \tau \end{bmatrix}. \tag{5.13}$$

The configurational partition function for that portion of an atactic vinyl polymer shown in Figure 5.5, which is a pentad stereosequence, becomes, according to Eq. 5.6,

$$Z = J^* U_l U_{ll} U_l U_{ld} U_d U_{dl} U_l U_{ld} U_d J. \tag{5.14}$$

Before we can calculate Z or any other conformation-dependent property of this vinyl polymer, the first-order $[\mu_1(\eta, \tau)]$ and second-order $[\mu_2(\omega, \omega', \omega'')]$ statistical weights must be determined. This is achieved by comparison of observed and calculated conformation-dependent properties, such as $\langle r^2 \rangle$, mean-square dipole moments $\langle u^2 \rangle$, and others, and adjusting the statistical weights until agreement is obtained. Matrix multiplication methods (Flory, 1969; Tonelli, 1986), which employ the RIS statistical weight matrices U_i, are utilized to calculate a variety of conformation-dependent properties of polymer chains.

5.2.2. Average Bond Conformations

Because the γ-*gauche* effect on ^{13}C NMR chemical shifts depends on the shielding produced by a γ-substituent in a *gauche* arrangement with the observed carbon nucleus, we must be able to determine bond conformational probabilities to implement the calculation of $\delta^{13}C$'s via the *γ-gauche-effect* method. This is a simple matter once we have determined the RIS model for a given polymer and expressed it in the form of statistical weight matrices for each of its constituent bonds. As an example, let us consider the atactic vinyl polymer fragment of Figure 5.5 and ask what is the probability of finding bond $i + 2$ in the *trans* conformation. The answer is simply the ratio $Z(\phi_{i+2} = t)/Z$, where Z is given by Equation 5.14 and $Z(\phi_{i+2} = t)$ is also obtained from Equation 5.14, but with the central statistical weight matrix corresponding to bond $i + 2$, $U_d(i + 2)$, replaced by

$$\begin{bmatrix} \eta & 0 & 0 \\ \eta & 0 & 0 \\ \eta & 0 & 0 \end{bmatrix}.$$

We have assigned the g^+ and g^- states of bond $i + 2$ statistical weights of 0 in order to obtain the probability $[P(\phi_{i+2} = t)]$ that this bond is in the *trans* conformation.

In Table 5.1 we present the probabilities of finding bond $i + 2$ in the *trans* conformation calculated as a function of the pentad stereosequence. (Note *ll* and *dd* diads are *m*, and *ld* and *dl* diads are *r*.) In the calculation it was assumed that $R = CH_3$, i.e., that the vinyl polymer in Figure 5.5 is polypropylene (PP). The RIS model developed for PP by Suter and Flory (1975) was used in the calculation. $P(\phi_{i+2} = t) = 0.44$–0.79, depending on the stereosequence of the pentad containing bond $i + 2$. This is a clear example of the

Table 5.1 ■ Calculated Probability of Finding Bond
ϕ in the *trans* Conformation

C	m,r	C	m,r	C	m,r	C	m,r	C
|		|		|$_\phi$		|		|
— C	— C	— C	— C	— C ⟩— C	— C	— C	— C	— C —

Pentad Stereosequence	$P(\phi = t)$
mrmr	.440
rrmr	.472
mmmm	.523
rmmr	.539
rmmm	.582
rrrr	.635
mrrm	.685
rrrm	.712
mmrr	.742
rmrm	.763
mmrm	.792

microstructural sensitivity of local bond conformations, which, via the γ-*gauche* effect, leads to the sensitivity of ^{13}C NMR chemical shifts to local polymer microstructure.

5.3. γ-*gauche*-Effect Calculation of ^{13}C NMR Chemical Shifts

5.3.1. Small-Molecule Example

We choose to describe the calculation of $\delta\,^{13}$C's for the three stereoisomeric forms of 2,4,6-trichloroheptane (TCH),

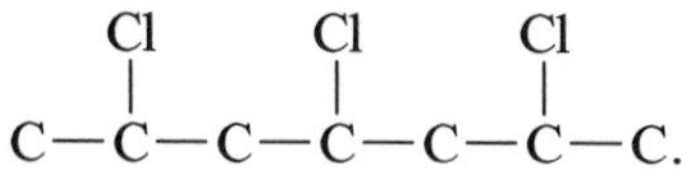

$$
\begin{array}{ccccccc}
 & Cl & & Cl & & Cl & \\
 & | & & | & & | & \\
C & -C & -C & -C & -C & -C & -C.
\end{array}
$$

The three-state (t, g^+, g^-) RIS model developed for poly(vinyl chloride) (PVC) by Williams and Flory (1969) and Flory and Pickles (1973) was used to calculate the bond rotation probabilities presented in Table 5.2. The PVC RIS model is characterized by the following first- and second-order interaction statistical weights: $\eta = 4.2$, $\tau = 0.45$, $\omega = \omega'' = 0.032$, and $\omega' = 0.071$, all appropriate to 25°C. These statistical weights were used to construct the matrices

Table 5.2 ▪ Calculated Rotation Probabilities for the Bonds in TCH

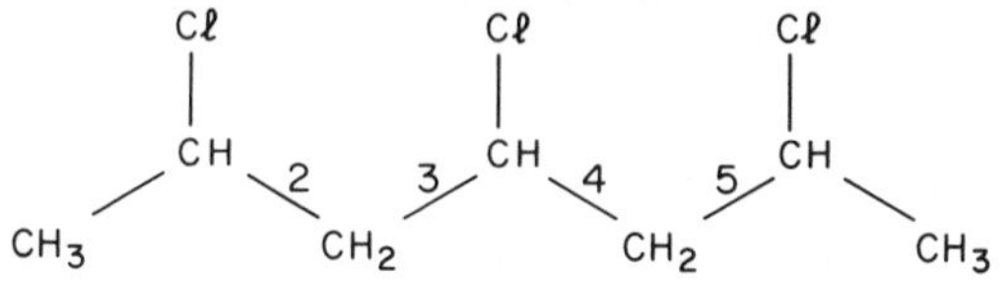

Bond	$P_{t, g, \bar{g}}$[a]		
	I	*S*	*H*
2	0.408	0.931	0.511
	0.571	0.065	0.471
	0.021	0.004	0.018
3	0.658	0.932	0.541
	0.320	0.063	0.441
	0.022	0.005	0.018
4	0.658	0.932	0.957
	0.320	0.063	0.039
	0.022	0.005	0.004
5	0.458	0.931	0.950
	0.521	0.065	0.046
	0.021	0.004	0.004

[a]$T = 25$°C.

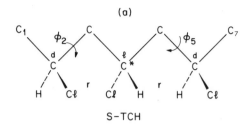

Figure 5.8 ■ (a) S-TCH in all-*trans*, planar zigzag conformation. (b), (c) Newman projections along bonds 2 and 5 of S-TCH.

of Equations 5.8–13, and bond conformational probabilities were evaluated for the *mm* or *I* (isotactic), *mr* (*rm*) or *H* (heterotactic), and *rr* or *S* (syndiotactic) stereoisomers of TCH.

Consider the central methine carbon (C*) in the *S*, or *rr* (*dld*), stereoisomer of TCH shown in Figure 5.8(a). We can see from the Newman projections in (b) and (c) that the *gauche* interactions of C* and its γ-substituents (C_1, C_7, Cl) may be written as follows: $\gamma_{C^*,C_1} = 1 - P(\phi_2 = t)$, $\gamma_{C^*,C_7} = 1 - P(\phi_5 = t)$, and $\gamma_{C^*,Cl} = 2 - P(\phi_2 = g^+) - P(\phi_5 = g^-)$. Consequently, the net summation of all γ-*gauche* interactions involving C* (γ_{C^*}) may be written as

$$\gamma_{C^*} = [2 - P(\phi_2 = t) - P(\phi_5 = t)] \times \gamma_{C^*, CH_3}$$
$$+ [2 - P(\phi_2 = g^+) - P(\phi_5 = g^-)] \times \gamma_{C^*, Cl}. \qquad (5.15)$$

Expressions similar to Equation 5.15 may also be written for γ_{C^*} in the *I* and *H* TCH isomers.

In Figure 5.9 we present the comparison of calculated and observed TCH δ ^{13}C's. Chemical shifts are presented in the form of stick spectra, where the farthest downfield resonance for each carbon type is assigned a 0.0-ppm chemical shift. Experimental spectra of the mixture of TCH stereoisomers were assigned (Tonelli et al., 1979) by observation of the 25-MHz spectra of

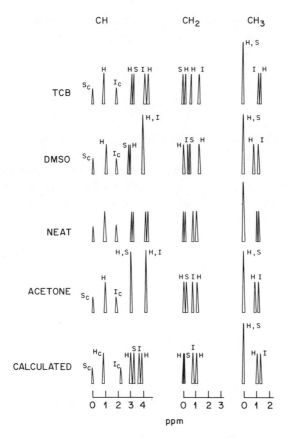

Figure 5.9 ■ Comparison of measured (33°C) and calculated ^{13}C chemical shifts of the TCH isomers in several solvents. The farthest downfield resonance in each of the three spectral regions is assigned a 0.0-ppm chemical shift. [Reprinted with permission from Tonelli et al. (1979).]

the pure stereoisomers. The following γ-*gauche* effects were used to obtain the calculated δ^{13}C's: $\gamma_{CH,CH_2 \text{ or } CH_3} = -5$ ppm, $\gamma_{CH_2 \text{ or } CH_3,CH} = -2.5$ ppm, and $\gamma_{CH,Cl} = -3$ ppm. Resonances denoted by the subscript c (S_c, H_c, I_c) correspond to the central methine carbons and appear downfield from the terminal methine resonances, because they have an additional β-substituent (CH) (see Section 4.2.2), which results in a deshielding of $+6.2$ ppm (Tonelli et al., 1979).

Note the close correspondence between the δ^{13}C's observed and calculated for the methine and methyl carbons, independent of the solvent used for measuring spectra. On the other hand, the δ^{13}C's calculated for the methylene carbons only agree with those measured in acetone. The sensitivity of the methylene carbon resonances to solvent is not the result of solvent-induced conformational changes, because in TCH the same bond rotations are involved in the γ-*gauche* effects of both the methylene and methine carbons, and yet only the methylene carbon resonances are sensitive to solvent. In fact, in 2,4-dichloropentane (DCP),

$$\begin{array}{ccc} \text{Cl} & & \text{Cl} \\ | & & | \\ \text{C}-\text{C}-\text{C}-\text{C}-\text{C}, \end{array}$$

the methylene carbon has no γ-substituents, yet it shows solvent-sensitive chemical shifts (Ando et al., 1977).

It is interesting that the shielding produced at the methyl carbon by a γ-*gauche* methine carbon, $\gamma_{\text{CH}_3,\text{CH}}$, is half ($-2.5$ ppm) that experienced by the methine carbon in a *gauche* arrangement with either CH_3 or CH_2, $\gamma_{\text{CH},\text{CH}_2 \text{ or } \text{CH}_3} = -5$ ppm. Based on our experience with the alkanes, both linear and branched (see Sections 4.2.3 and 4.2.4), we would have expected $\gamma_{\text{CH}_3,\text{CH}} = \gamma_{\text{CH},\text{CH}_2 \text{ or } \text{CH}_3} = -5$ ppm. Apparently the chlorine atom attached to the methine carbon reduces the shielding it produces at the methyl carbon when they are in a γ-*gauche* arrangement.

5.3.2. Macromolecular Example

We may now attempt to calculate the stereosequence-dependent $\delta^{13}C$'s observed in the ^{13}C NMR spectra of atactic PVC (Tonelli et al., 1979). γ-effects derived by comparison of observed and calculated $\delta^{13}C$'s for the PVC model compound TCH, i.e., $\gamma_{\text{CH}_2,\text{CH}} = -2.5$ ppm, $\gamma_{\text{CH},\text{CH}_2} = -5$ ppm, and $\gamma_{\text{CH},\text{Cl}} = -3$ ppm, are assumed to be applicable to PVC. For the methine carbons, bond conformation probabilities were calculated for all pentad stereosequences:

$$\begin{array}{ccccccccc} \text{Cl} & \text{m,r} & \text{Cl} & \text{m,r} & \text{Cl} & \text{m,r} & \text{Cl} & \text{m,r} & \text{Cl} \\ | & & | & \phi_1 & | & & \phi_2 & | & | \\ -\text{c}-\text{c}-\text{c}\rightarrow\text{c}-\text{c}^*-\text{c}\rightarrow\text{c}-\text{c}-\text{c}- \end{array}$$

Only the bond rotations ϕ_1 and ϕ_2 effecting the γ-*gauche* arrangements of the central methine carbon (C*) were considered. All tetrad stereosequences

$$\begin{array}{ccccccc} \text{Cl} & \text{m,r} & \text{Cl} & \text{m,r} & \text{Cl} & \text{m,r} & \text{Cl} \\ | & \phi_a & | & & | & \phi_b & | \\ -\text{c}-\text{c}\rightarrow\text{c}-\text{c}^+-\text{c}\rightarrow\text{c}-\text{c}- \end{array}$$

were considered when evaluating the rotational state probabilities for ϕ_a and ϕ_b, which govern the γ-gauche interactions of the central methylene carbon (C$^+$). The RIS model developed for PVC (Williams and Flory, 1969; Flory and Pickles, 1973) was used to calculate the bond conformation probabilities.

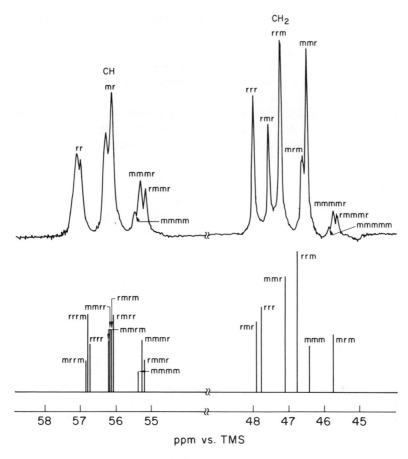

Figure 5.10 ■ Comparison of measured ¹³C NMR spectrum of atactic PVC dissolved in TCB at 120°C with the chemical shifts calculated from evaluation of γ-effects. [Reprinted with permission from Tonelli et al. (1979).]

In Figure 5.10 the ¹³C NMR spectrum of atactic PVC observed at 90.5 MHz is compared with the stick spectrum of calculated δ¹³C's. The assignments indicated in the observed spectrum were obtained independently (Carman, 1973) from the polymerization statistics of PVC. (See Chapter 6 for an explanation of this method of resonance assignment.) The correspondence between calculated and observed chemical shifts in the methine region is close, and it confirms the γ-effect model of ¹³C chemical shifts in PVC derived from the ¹³C NMR study of TCH isomers.

On the other hand, the methylene portion of the spectrum obtained in 100% TCB (1,2,4-trichlorobenzene) differs markedly from the predicted δ¹³C's. The solvent dependence of the methylene carbon region in ¹³C NMR spectra of

PVC, which had been noted previously (Ando et al., 1976), is not unexpected in view of the results of studies of the PVC model compounds TCH (Tonelli et al., 1979) and DCP (Ando, et al., 1977). In fact we (Schilling and Tonelli, 1979) have been able to reproduce experimentally the methylene portion of the calculated spectrum by performing the measurements in a mixed solvent of TCB and dimethylsulfoxide.

Having presented examples of the γ-*gauche*-effect method for calculating ^{13}C NMR chemical shifts, we now turn to their application in the determination of polymer microstructures in the remaining chapters of this book.

References

Abe, A., Jernigan, R. L., and Flory, P. J. (1966). *J. Am. Chem. Soc.* **88**, 631.

Ando, I., Kato, Y., and Nishioka, A. (1976). *Makromol. Chem.* **177**, 2759.

Ando, I., Kato, Y., Kondo, M., and Nishioka, A. (1977). *Makromol. Chem.* **178**, 803.

Bartell, L. S. and Kohl, D. A. (1963). *J. Chem. Phys.* **39**, 3097.

Birshstein, T. M. and Ptitsyn, O. B. (1966). *Conformation of Macromolecules*, translated from the Russian by S. N. Timasheff and M. J. Timasheff, Wiley-Interscience, New York.

Bonham, R. A., and Bartell, L. S. (1959). *J. Am. Chem. Soc.* **81**, 3491.

Borisova, N. P. (1964). *Vysokomol. Soedin.* **6**, 135.

Bovey, F. A. (1982). *Chain Structure and Conformation of Macromolecules*, Academic Press, New York, Chapter 7.

Carman, C. J. (1973). *Macromolecules* **6**, 725.

Flory, P. J. (1969). *Statistical Mechanics of Chain Molecules*, Wiley-Interscience, New York.

Flory, P. J. and Pickles, C. J. (1973). *J. Chem. Soc. Faraday Trans. 2* **69**, 632.

Hendrickson, J. B. (1961). *J. Am. Chem. Soc.* **83**, 4537.

Herschback, D. R. (1963). *International Symposium on Molecular Structure and Spectroscopy, Tokyo, 1962*, Butterworths, London.

Ising, E. (1925). *Z. Phys.* **31**, 253.

Kramers, H. A. and Wannier, G. H. (1941). *Phys. Rev.* **60**, 252.

Kuchitsu, K. (1959). *J. Chem. Soc. Jpn.* **32**, 748.

Mansfield, M. L. (1983). *Macromolecules* **16**, 1863.

Mizushima, S. (1954). *Structure of Molecules and Internal Rotation*, Academic Press, New York.

Newell, G. F., and Montroll, E. W. (1953). *Rev. Mod. Phys.* **25**, 353.

Schilling, F. C. and Tonelli, A. E. (1979). Unpublished observations.

Scott, R. A. and Scheraga, H. A. (1966). *J. Chem. Phys.* **44**, 3054.

Suter, U. W. and Flory, P. J. (1975). *Macromolecules* **8**, 765.

Tonelli, A. E. (1986). *Encyclopedia of Polymer Science and Engineering*, Second Ed., Wiley, New York, Vol. 4, p. 120.

Tonelli, A. E., Schilling, F. C., Starnes, W. H., Jr., Shepherd, L., and Plitz, I. M. (1979). *Macromolecules* **12**, 78.

Volkenstein, M. V. (1963). *Configurational Statistics of Polymeric Chains*, translated from the Russian by S. N. Timasheff and M. J. Timasheff, Wiley-Interscience, New York.

Williams, A. D. and Flory, P. J. (1969). *J. Am. Chem. Soc.* **91**, 3118.

Wilson, E. B., Jr. (1959). *Adv. Chem. Phys.* **2**, 367.

Wilson, E. B., Jr. (1962). *Pure Appl. Chem.* **4**, 1.

6

Determination of Stereosequences in Vinyl Polymers

6.1. Introduction

Our task in this chapter is to describe how NMR spectroscopy is utilized to determine the configurational stereosequences in vinyl homopolymers

$$\begin{array}{c} R \\ | \\ +CH-CH_2+, \end{array}$$ where R can be methyl, phenyl, acetate, OH, Cl, CN, etc.

We seek to establish the sequence of configurations at consecutive methine carbons (CH). Is the vinyl polymer stereoregular, i.e., isotactic with virtually all monomer units of the same configuration or syndiotactic with an alternating succession of monomer configurations, or is it atactic with some statistical distribution of monomer stereosequences? As noted in Chapter 1, the impetus for these questions is the fact that the physical properties of vinyl polymers are intimately related to their microstructures, among which the monomer stereosequence stands out as the most important.

We adopt the m, r diad nomenclature of Bovey (1969) to describe the stereosequence of vinyl polymers. Adjacent monomer pairs (diads) with the same relative configurations are called m (meso) diads, while those with opposite configurations are termed r (racemic) diads. Because of symmetry, the chemical-shift sensitivity of a carbon nucleus to configurational-sequence lengths varies according to its position in the monomer unit. Methylene carbons are directly bonded to two asymmetric centers and may therefore show configurational sensitivity to diads, tetrads, hexads, etc., as illustrated in Figure 6.1(a). For methine and the attached side-chain carbons (R), the asymmetric carbons are next nearest neighbors, and their sensitivity to configurational sequences begins with triads, and proceeds to pentads, heptads, etc., as shown in part (b).

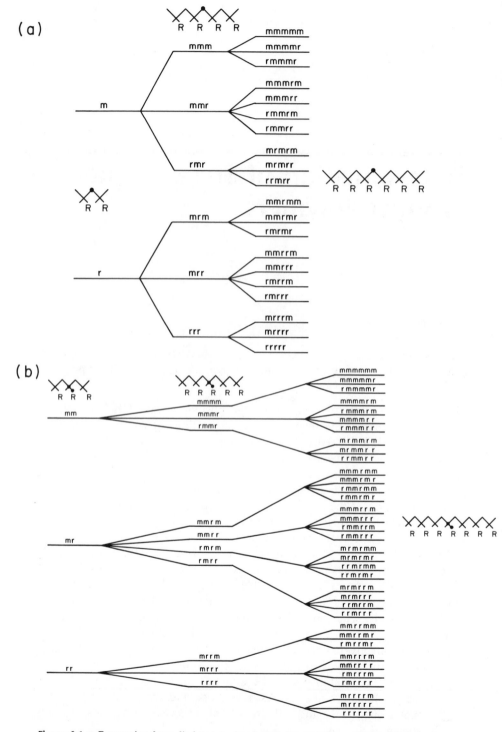

Figure 6.1 ■ Progression from diad to tetrad to hexad stereosequences for methylene carbons (a) and from triad to pentad to heptad stereosequences for methine and side-chain carbons (b) in vinyl polymers. [Reprinted with permission from Randall (1977).]

Let us consider the number of unique configurations possible for each stereosequence length. This is most readily determined by introducing (Price, 1962) 0 and 1 (or d and l) to distinguish between the two possible monomer configurations. On this basis there are four possible diads 00, 11, 01, and 10, but only two, $m = 00 = 11$ and $r = 01 = 10$, are unique, because the methine carbons are not true asymmetric centers. Isolated 0 and 1 centers cannot be distinguished; only relative diad configurations, such as 00(11) and 01(10), are unique. Of the eight (2^3) possible triads, only three are distinguishable: $mm = 111 = 000$, $mr = rm = 110 = 001 = 011 = 100$, and $rr = 101 = 010$. Similarly, only six of the possible sixteen (2^4) tetrads are unique: $mmm = 1111 = 0000$, $mmr = rmm = 1110 = 0001 = 1000 = 0111$, $rmr = 1001 = 0110$, $mrm = 1100 = 0011$, $mrr = rrm = 1101 = 0010 = 1011 = 0100$, and $rrr = 1010 = 0101$.

In general (Frisch et al., 1966) the number $N(n)$ of observationally distinguishable types of sequences containing n monomer units is given by

$$N(n) = 2^{n-2} + 2^{m-1}, \qquad (6.1)$$

where $m = n/2$ for even n and $m = (n - 1)/2$ for odd n. We therefore expect to be able to observe as many as $2, 3, 6, 10, 20, 36$, etc. resonances for the carbon nuclei in vinyl polymers when their ^{13}C chemical shifts show sensitivities to diads, triads, tetrads, pentads, hexads, heptads, and longer stereosequences (see Figure 6.1).

Based on elementary statistical considerations (Frisch et al., 1966), necessary relationships may be drawn to connect the probabilities of occurrence of the various observed stereosequences. Several of these relationships are presented in Table 6.1. These relationships are completely general and do not depend on the configurational statistics of monomer addition produced by any particular polymerization mechanism. Consequently, they serve as useful tests of peak assignments in the NMR spectra of vinyl polymers. If any set of assignments results in peak intensity ratios in conflict with the appropriate relationship in Table 6.1, then this assignment cannot be correct.

We now turn to a discussion of how NMR spectroscopy is utilized to determine the stereosequence of vinyl polymers. This discussion is divided into three separate approaches: traditional methods, 2D NMR techniques, and the γ-gauche-effect method. Traditional methods, which include the NMR observation of model compounds and stereoregular polymers, epimerization studies, and inference of stereosequences from an assumed polymerization mechanism, will only be briefly discussed. Randall (1977) has thoroughly covered these approaches for determining the stereosequences of vinyl polymers, and the reader is referred to Randall's book for further details and the many examples he discusses.

The more recently developed 2D NMR and γ-gauche-effect methods for determining vinyl polymer stereosequences are presented in greater detail. Emphasis is placed on advantages over the more traditional methods, which they bring to the process of stereosequence determination. Both of these newer

Table 6.1 ■ Some Necessary Relationships among Sequence Frequencies

Diad–diad	$(m) + (r) = 1$
Triad–triad	$(mm) + (mr) + (rr) = 1$
Tetrad–tetrad	sum $= 1$ $(mmr) + 2(rmr) = 2(mrm) + (mrr)$
Pentad–pentad	sum $= 1$ $(mmmr) + 2(rmmr) = (mmrm) + (mmrr)$ $(mrrr) + 2(mrrm) = S(rrmr) + (rrmm)$
Diad–triad	$(m) = (mm) + \frac{1}{2}(mr)$ $(r) = (rr) + \frac{1}{2}(mr)$
Diad–tetrad	$(m) = (mmm) + (mrm) + \frac{1}{2}(mmr) + \frac{1}{2}(mrr)$ $(r) = (rrr) + (rmr) + \frac{1}{2}(mmr) + (mrr)$
Diad–pentad	$(m) = (mmmm) + (mmmr) + (rmmr)$ $\qquad + \frac{1}{2}(mmrm) + \frac{1}{2}(mmrr) + \frac{1}{2}(mrmr) + \frac{1}{2}(rmrr)$ $(r) = (rrrr) + (mrrr) + (mrrm)$ $\qquad + \frac{1}{2}(mmrm) + \frac{1}{2}(mmrr) + \frac{1}{2}(mrmr) + \frac{1}{2}(rmrr)$
Triad–tetrad	$(mm) = (mmmm) + \frac{1}{2}(mmmr)$ $(mr) = (mmr) + 2(rmr) = (mrr) + 2(mrm)$ $(rr) = (rrr) + \frac{1}{2}(mrr)$
Triad–pentad	$(mm) = (mmmm) + (mmmr) + (rmmr)$ $(mr) = (mmrm) + (mmrr) + (mrmr) + (rmrr)$ $(rr) = (rrrr) + (mrrr) + (mrrm)$
Tetrad–pentad	$(mmmm) = (mmmm) + \frac{1}{2}(mmmr)$ $(mmr) = (mmmr) + 2(rmmr) = (mmrm) + (mmrr)$ $(rmr) = \frac{1}{2}(mrmr) + \frac{1}{2}(rmrr)$ $(mrm) = \frac{1}{2}(mrmr) + \frac{1}{2}(mmrm)$ $(rrm) = 2(mrrm) + (mrrr) = (mmrr) + (rmrr)$ $(rrr) = (rrrr) + \frac{1}{2}(mrrr)$

methods are free of the assumptions and eliminate the need for model compound studies which have often been integral to the earlier NMR studies of vinyl polymer stereosequence. These advantages originate from the physical basis of each method. The γ-*gauche*-effect method relies on knowledge of the dependence of local vinyl polymer conformation upon stereosequence, while the 2D NMR method (COSY) provides direct observation of the local microstructural connectivity in a vinyl polymer. These two features are stressed in the remainder of this chapter, and serve to recommend both methods as superior means to learn about the stereosequences present in vinyl polymers.

6.2. Traditional Methods

6.2.1. Stereoregular Polymers

We choose to use polypropylene (PP) as the subject of our discussion of traditional approaches to the NMR determination of vinyl polymer stereosequence. Several reasons influenced this selection. First, both stereoregular forms of PP can be synthesized, i.e., isotactic (*i*-PP) and syndiotactic (*s*-PP). Second, several PP model compounds have been made and studied by ^{13}C NMR. Third, stereoregular PP can be epimerized to obtain an atactic material (*a*-PP), with an equilibrium (Bernoullian or random) distribution of stereosequences. Fourth, PP may be polymerized in several different ways to obtain *a*-PPs with widely different stereosequence distributions. For these reasons a complete determination of PP stereosequences to the pentad level is possible via the traditional methods of NMR stereosequence analysis. In this regard PP is unique among vinyl polymers. With the recent announcement of

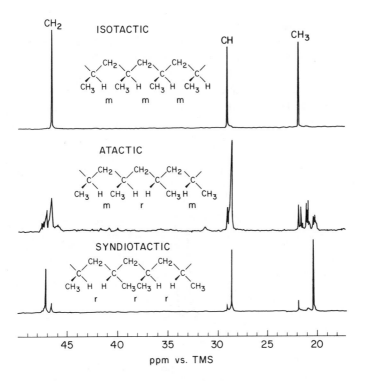

Figure 6.2 ■ ^{13}C NMR spectra at 25 MHz for PPs dissolved (20% w/v) in 1,2,4-trichlorobenzene at 140°C. Schematic representation illustrates *m* (meso) and *r* (racemic) dyads and polymer chain tacticity: isotactic (...*mmmmm*...), syndiotactic (...*rrrrr*...), atactic or heterotactic (...*mmrmrrmr*...). [Reprinted with permission from Tonelli and Schilling (1981).]

the polymerization of highly syndiotactic polystyrene (Ishihara et al., 1986), polystyrene joins PP to become one of only two vinyl polymers capable of being produced in both stereoregular forms, which are susceptible to epimerization.

The ^{13}C NMR spectra recorded at 25 MHz for three PP samples (Tonelli and Schilling, 1981) are presented in Figure 6.2. Assignment of resonances according to carbon type (CH, CH$_2$, and CH$_3$) are achieved through utilization of the α, β, γ-substituent effects on δ^{13}C's described previously (see Section 4.2.2) or by off-resonance decoupling (Randall, 1977) of the ^{1}H nuclei. Off-resonance ^{1}H decoupling is performed by offsetting the ^{1}H decoupling frequency some 100–200 Hz from the resonance of interest instead of applying broad-band ^{1}H noise decoupling, which removes all ^{13}C–^{1}H splittings. ^{13}C splittings produced by off-resonance ^{1}H decoupling are narrow (a few hertz) and usually do not result in overlap with the resonances of other carbon types. Because both ^{13}C and ^{1}H are spin-$\frac{1}{2}$ nuclei, and their resonant frequencies are widely separated; the $n + 1$ rule (Roberts, 1959) is obeyed for the ^{13}C–^{1}H splitting patterns. Thus, the number of ^{13}C resonances expected for each carbon type in an off-resonance ^{1}H-decoupled spectrum is simply $n + 1$, or the number of directly bonded protons plus one.

Comparison of the resonances observed in the ^{13}C NMR spectra of stereoregular i-PP and s-PP to those recorded for a-PP permits the identification of two resonances in each region of the a-PP spectrum. This identification is made in the methyl carbon region of the ^{13}C NMR spectrum recorded for a-PP at 90.5 MHz and presented in Figure 6.3. Though the comparison of resonances observed for stereoregular PPs provides a point of departure in the task of unravelling the stereosequence of a-PP, more than twenty resonances remain to be assigned in the methyl region of its spectrum. Because over twenty resonances are observed, we can conclude that δ^{13}C's of the methyl

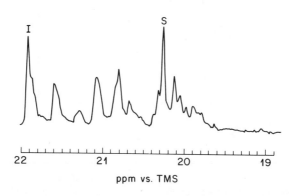

Figure 6.3 ■ ^{13}C NMR spectrum at 90.52 MHz of the methyl carbon region in atactic PP in 20% w/v n-heptane solution at 67°C. [Reprinted with permission from Tonelli and Schilling (1981).]

carbons in *a*-PP are sensitive to heptad stereosequences (see Figure 6.3 and Table 6.1). Of the 36 possible heptad stereosequences we now are able to identify two, i.e., $I = mmmmmm$ and $S = rrrrrr$.

6.2.2. Epimerization of Stereoregular Polymers

With the use of appropriate catalysts several vinyl polymers have shown the ability to undergo configurational inversions at the pseudoasymmetric methine carbons (Stehling and Knox, 1975; Shepherd et al., 1979; Suter and Neuenschwander, 1981; Dworak et al., 1985). Stehling and Knox (1975) found that addition of 1% dicumylperoxide plus 4% tris(2,3-dibromopropyl)phosphate to *i*-PP and *s*-PP resulted in random, well-spaced inversions of monomer unit configurations at low degrees of conversion. Epimerization of *i*-PP results in the introduction of the following pentad stereosequences:

0	0	0	0	0	0	0	0	0	0	0
					↓		epimerization			
0	0	0	0	0	1	0	0	0	0	0
					↓					
0	0	0	0	1	(2)		*mmmr*			
0	0	0	1	0	(2)		*mmrr*			
0	0	1	0	0	(1)		*mrrm*			

In the case of *s*-PP epimerization leads to

0	1	0	1	0	1	0	1	0	1	0
					↓					
0	1	0	1	0	0	0	1	0	1	0
					↓					
0	1	0	0	0	(2)		*rrmm*			
1	0	1	0	0	(2)		*rrrm*			
1	0	0	0	1	(1)		*rmmr*			

Thus epimerization of stereoregular PPs produces three new resonances in a $2:2:1$ ratio in the methyl region of the ^{13}C NMR spectra of *i*-PP and *s*-PP. Because in each case the *rmmr* and *mrrm* resonances appear at half the intensity of the other two introduced pentad resonances, they are readily assigned. The *mmrr* resonance is identified by comparison of the spectra of epimerized *i*-PP and *s*-PP, since it is the only commonly produced resonance and appears in both spectra. The *mrrr* and *rmmm* resonances are then assigned by default. At this stage we are able to assign seven of the ten possible pentad stereosequences, i.e., *mmmm* and *rrrr* from *i*-PP and *s*-PP, and *mmmr*, *mmrr*, *mrrm*, *rrrm*, and *rmmr* from epimerization of *i*-PP and *s*-PP.

The remaining three pentad stereosequences *mmrm*, *rmrr*, and *rmrm* may be assigned by consideration of the necessary pentad–pentad relations (see Table 6.1):

$$2\,rmmr + mmmr = mmrm + mmrr, \qquad (6.2)$$

$$2\,mrrm + mrrr = mmrr + rmrr. \qquad (6.3)$$

This assignment would be based on discriminating differences in the intensities observed for $mmrm$, $rmrr$, and $rmrm$ pentads and lead to the tentative assignment of $rmrr$, $mmrm$, and $rmrm$ from low to high field.

6.2.3. Model Compounds

A "tour de force" synthesis and ^{13}C NMR study of ^{13}C-labeled PP heptad model compounds by Zambelli et al. (1975) confirmed and clarified the assignments made by comparison of the spectra of i-PP and s-PP before and after epimerization. $3(s), 5(r), 7(rs), 9(rs), 11(rs), 13(r), 15(s)$-heptamethyl-heptadecane (A) and a mixture of A with $3(s), 5(s), 7(rs), 9(rs), 11(rs),$ $13(r), 15(s)$-heptamethylheptadecane, all ^{13}C-enriched at the 9-CH$_3$ position, permitted the assignment of nine pentad stereosequences in a-PP. When the results obtained from the ^{13}C NMR spectra recorded for the stereoregular PPs before and following epimerization and for the PP heptad model compounds of Zambelli et al. (1975) are combined, the following pentad stereosequence assignment results for a-PP: from low to high field, $mmmm$, $mmmr$, $rmmr$, $mmrr$, $mmrm$, $rmrr$, $rmrm$, $mrrm$, $mrrr$, and $rrrr$.

6.2.4. Assumed Polymerization Mechanism

^{13}C NMR chemical-shift assignments in vinyl polymer spectra are often aided by comparing the intensities of the observed stereosequence resonances with those expected from a statistical model of the polymerization mechanism. Two such models of vinyl polymerization are represented schematically in Figure 6.4. (Bovey, 1972). In a "random addition" of monomers, or a Bernoullian polymerization, the probability P_m of adding a monomer to form a meso or m diad is independent of the configurations of the previously added monomer units. As a consequence, the probability of forming an r diad is $P_r = 1 - P_m$. A vinyl polymer that has a Bernoullian distribution of stereosequences is configurationally random only when $P_m = P_r = 0.5$. The fractional populations of the various stereosequences may be directly obtained from P_m or P_r. As examples, $m = P_m$, $r = 1 - P_m$, $mm = P_m^2$, $mr = 2P_m(1 - P_m)$, $rr = (1 - P_m)^2$, $mrm = P_m(1 - P_m)P_m = P_m^2(1 - P_m)$, and $rmmr = (1 - P_m)P_m P_m(1 - P_m) = P_m^2(1 - P_m)^2$. To test the consistency of a proposed ^{13}C NMR resonance assignment for a vinyl polymer against a Bernoullian distribution of stereosequences, we must compare the observed intensities of the assigned peaks with those predicted from a Bernoulli trial.

   In the previous chapter (see Section 5.3.2), where we discussed the ^{13}C NMR spectra of poly(vinyl chloride) (PVC), it was mentioned that

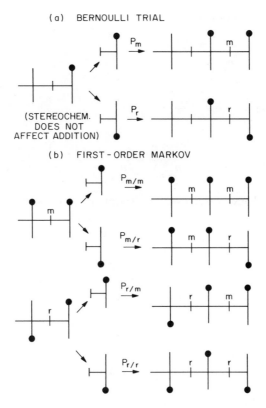

Figure 6.4 ■ Schematic representation of Bernoulli trial and first-order Markov propagation steps. [Reprinted with permission from Bovey (1972).]

Carman (1973) had assigned the spectrum by comparison of observed resonance intensities with those predicted by the polymerization statistics. He found both the methine and methylene carbon regions of the spectrum to be described by a Bernoullian distribution of stereosequences with $P_m = 0.45$, not far from a completely random distribution.

If during a polymerization the configuration of an added monomer depends upon the configurations of previously added units, then the resulting stereosequence is governed by Markov statistics [see Figure 6.4(b)]. A first-order Markov polymerization is depicted in Figure 6.4(b), where only the configuration of the previous diad influences the formation of the next diad added. This polymerization scheme is characterized by the conditional probabilities $P_{m/r}$, $P_{m/m}$, $P_{r/m}$, and $P_{r/r}$, where for example $P_{r/m}$ is the probability of adding an m diad to an r chain end. Of course $P_{m/r} + P_{m/m} = P_{r/m} + P_{r/r} = 1.0$. Stereosequence populations may be written (Bovey, 1972) in terms of these conditional probabilities, such as

$$mm = \frac{P_{m/m} + P_{r/m}}{P_{m/r} + P_{r/m}}, \tag{6.4}$$

$$mrm = \frac{P_{m/r}P_{r/m}^2}{P_{m/r} + P_{r/m}}, \tag{6.5}$$

$$mrmm = \frac{2P_{m/r}P_{r/m}^2(1 - P_{m/r})}{P_{m/r}(1 - P_{r/m}) + 2P_{m/r}P_{r/m} + P_{r/m}(1 - P_{m/r})}. \tag{6.6}$$

Just as the frequency of stereosequences (intensities) observed in the ^{13}C NMR spectrum of PVC were fit to those predicted by a Bernoulli trial, other polymers, most notably those polymerized by ionic initiation, often have stereosequences described by first-order Markov statistics (Randall, 1977). For those atactic vinyl polymers that cannot be obtained in stereoregular forms, or where appropriate model compounds have not been synthesized, comparison of the observed resonance intensities with the statistical predictions of a polymerization model can sometimes materially aid the analysis of their stereosequence.

6.3. 2D NMR Determination of Vinyl Polymer Stereosequence

Since it was first conceived by Jeener in 1971, two-dimensional J-correlated spectroscopy (COSY) has been extensively applied to resonance assignments in the ^{1}H NMR spectra of proteins, peptides, nucleic acids, and synthetic polymers [see for example Wüthrich (1986) and Bovey and Mirau (1988)]. This is a consequence of the ability of 2D COSY spectroscopy to connect spins that are J-coupled to each other yielding spectra that are maps of the entire coupling network within molecules (see Section 5 in Chapter 3). Here we discuss a single application of 2D COSY spectroscopy used to determine the stereosequence of a vinyl polymer.

Poly(vinyl fluoride) (PVF) $+CH_2-CHF+$, as produced commercially, is a highly crystalline, clear plastic that contains a substantial fraction (ca. 10%) of head-to-head : tail-to-tail monomer units [Wilson and Santee (1965); Tonelli et al. (1982)]. Cais and Kometani (1984) were able to produce isoregic PVF, completely free of inverted monomer units, by reductively dechlorinating a precursor polymer $+CH_2-CFCl+$ with tri-n-butyltin hydride. Bruch et al. (1984) applied the 2D COSY experiment to the ^{19}F NMR spectrum of isoregic PVF. ^{19}F NMR spectra, like their ^{13}C NMR counterparts, are much more sensitive to polymer microstructure than are ^{1}H NMR spectra (see Section 3.6).

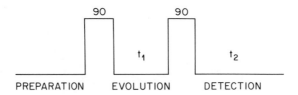

Figure 6.5 ▪ 2D J-correlated pulse sequence.

The pulse sequence for the 2D COSY experiment is shown in simplified form in Figure 6.5 (see also Figure 3.9). The broadband proton decoupler was kept on during the entire sequence to remove the extensive ^{19}F-^{1}H scalar J-coupling. Free induction decays detected during t_2 following systematic incrementation of the evolution time t_1 yield a data matrix that is Fourier transformed in both dimensions to give a 2D spectrum as a function of the two precession frequencies f_1 and f_2, during t_1 and t_2, respectively. Those fluorines not exchanging magnetization have $f_1 = f_2$ and appear in the normal spectrum along the diagonal corresponding to $f_1 = f_2$. Fluorine nuclei that do exchange magnetization via J-coupling have final frequencies different from their initial precession frequencies, or $f_1 \neq f_2$. These coupled fluorines yield off-diagonal peaks. The COSY spectrum of two J-coupled fluorine nuclei with normal precession frequencies f_a and f_b consists of two diagonal peaks at (f_a, f_a) and (f_b, f_b) and two off-diagonal peaks at (f_a, f_b) and (f_b, f_a). Though reduced in intensity compared to the diagonal peaks, the off-diagonal or cross peaks contain the useful information in a 2D COSY spectrum. By matching all pairs of cross peaks, it is possible to establish the network of spin connectivities via J-coupling.

The ^{19}F NMR spectrum of isoregic PVF measured at 188 MHz (Bruch et al., 1984) is presented in Figure 6.6. Triad stereosequences mm, $mr(rm)$, and rr were assigned previously by Weigert (1971) by analogy with the ^{13}C NMR spectrum of atactic polypropylene and by Tonelli et al. (1982) using the γ-*gauche*-effect method of predicting ^{19}F NMR chemical shifts (see Chapter 7). Because the relative intensities of the resonances indicate a nearly random atactic, Bernoullian stereosequence, with $P_m = 0.48$, pentad assignments are difficult to make on this basis. However, Bruch et al. (1984) were able to make all ten pentad assignments through utilization of ^{19}F 2D COSY spectroscopy.

Four-bond J-coupling (about 7 Hz) between neighboring fluorines, though too small to be resolved in the normal ^{19}F NMR spectrum (Figure 6.6) with line widths of about 14 Hz, is large enough to give rise to cross peaks in the COSY spectrum. Because there is expected to be coupling between the central pair of fluorines in pentad sequences sharing a common hexad, these cross peaks can be used to make stereosequence assignments. As an example, in the hexad sequence

$$- \overset{\overset{\text{H}}{|}}{\underset{\underset{\text{F}}{|}}{\text{C}}} - \text{CH}_2 - \overset{\overset{\text{F}}{|}}{\underset{\underset{\text{H}}{|}}{\text{C}}} - \text{CH}_2 - \overset{\overset{\text{F}^*}{|}}{\underset{\underset{\text{H}}{|}}{\text{C}}} - \text{CH}_2 - \overset{\overset{\text{F}^*}{|}}{\underset{\underset{\text{H}}{|}}{\text{C}}} - \text{CH}_2 - \overset{\overset{\text{H}}{|}}{\underset{\underset{\text{F}}{|}}{\text{C}}} - \text{CH}_2 - \overset{\overset{\text{H}}{|}}{\underset{\underset{\text{F}}{|}}{\text{C}}} - $$

four-bond coupling between the indicated fluorines will result in cross peaks between the *rmmr* and *mmrm* pentad resonances in the COSY spectrum. All ten pentad assignments can be made unambiguously from such *J*-coupling correlations.

The ^{19}F 2D COSY spectrum of isoregic PVF is presented in Figure 6.7. Cais and Komentani (1984) have shown that the three upfield resonances correspond to the *rr*-centered pentads. Without this independent assignment,

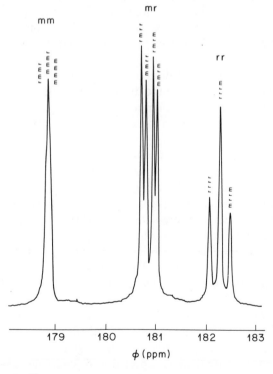

Figure 6.6 ■ 188-MHz ^{19}F spectrum of an 11% solution of isoregic poly(vinyl fluoride) in DMF-d_7 at 130°C. Broad-band proton decoupling was employed to remove ^{19}F–^{1}H coupling. [Reprinted with permission from Bruch et al. (1984).] ϕ measured vs. CFCl$_3$.

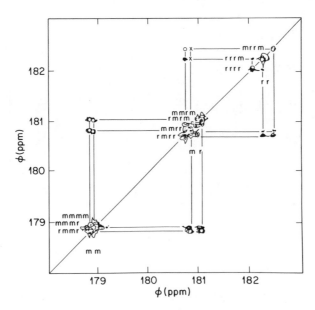

Figure 6.7 ■ 188-MHz [19]F 2D *J*-correlated spectrum of an 11% solution of isoregic poly(vinyl fluoride) in DMF-d_7 at 130°C. [Reprinted with permission from Bruch et al. (1984).]

it would be difficult to choose between the *rr*- and *mm*-centered pentads, because $P_m = 0.48$ for this PVF. Because the *rrrr* pentad can couple only to the *rrrm* pentad, it is easily identified, and consequently *rrrm* must be the central peak. By default the *mrrm* resonance must be farthest upfield.

In the central *mr*-centered region only one pentad *mmrr* is expected to be coupled to both the *rr*- and *mm*-centered regions via the *rrrm*, *mrrm* and *mmmr*, *rmmr* pentads, respectively. This is the second farthest downfield *mr* resonance. The only other resonance that can couple to the *rr* region is *rmrr*, and it must therefore be the farthest downfield peak. Similarly, *mmrm* is the farthest upfield resonance, because, aside from *mmrr*, it is the only other peak that can be coupled to the *mm* region. By default, the second farthest upfield resonance is *rmrm*, and, as expected, it is only coupled to the *mr* region.

In the normal ^{19}F NMR spectrum (Figure 6.6), all *mm*-centered pentads appear to have the same chemical shifts. However, in the COSY spectrum *mm*-centered-pentad fine structure is revealed. A slice through the *mm* region of Figure 6.7 is shown in Figure 6.8, where three peaks along the dashed diagonal are evident. Only the farthest upfield peak is uncoupled to the *mr* region and is therefore assigned to the *mmmm* pentad. Based on relative intensities, i.e., *mmmr* : *rmmr* = 2 : 1, *mmmr* must be the central peak and *rmmr* the farthest downfield resonance.

Aside from the ambiguity between *mm*- and *rr*-centered resonances found in this nearly random PVF ($P_m = 0.48$), all pentad resonance assignments were

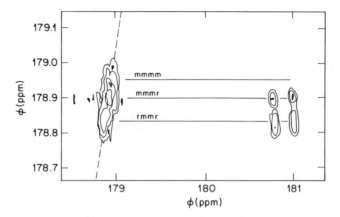

Figure 6.8 ▪ Expansion of the *mm* region of the 2D *J*-correlated spectrum of isoregic poly(vinyl fluoride) shown in Figure 6.7. [Reprinted with permission from Bruch et al. (1984).]

made in a single 2D COSY experiment. This was achieved without the synthesis of PVF model compounds or stereoregular PVFs, and vividly illustrates the power of 2D NMR in the determination of polymer microstructure. In addition, there now exist ^{1}H–^{13}C heteronuclear shift-correlated 2D NMR techniques (Bax, 1983) which permit unambiguous assignment of resonances belonging to *m* and *r* diads in vinyl polymers. The success of these techniques rests on the observation that the methylene protons in an *m*-diad are magnetically nonequivalent, while those belonging to an *r*-diad are degenerate (see Section 3.2). Chang et al. (1985) have recently applied these techniques to identify the ^{1}H and ^{13}C resonances belonging to the *m* and *r* diads in poly(vinylamine).

6.4. Application of γ-*gauche*-Effect Method

We conclude the description of NMR techniques used to determine the stereosequences of vinyl polymers by presenting an example of the γ-*gauche*-effect method for predicting and thereby assigning ^{13}C chemical shifts. Mention has been made (Section 6.2.1, Figure 6.3) that the ^{13}C NMR spectra of atactic polypropylene (*a*-PP) shows sensitivity to heptad stereosequences in the methyl region. In Figure 6.9 an *a*-PP heptad is illustrated along with Newman projections detailing the γ-*gauche* interactions involving the methyl group. It is clear that in the *t* and *g*⁻ backbone conformations the methyl group is gauche to its γ-substituents, the backbone methine carbons (α). To predict the ^{13}C chemical shifts expected for the methyl carbons in *a*-PP we simply have to calculate the *trans* and *gauche* probabilities for the backbone bonds in each of the 36 heptad stereosequences. When this is carried out with

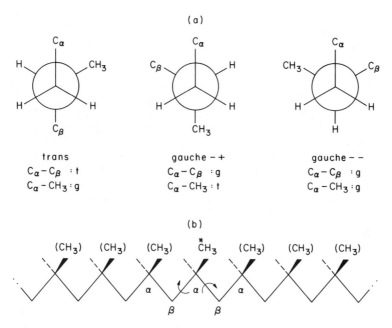

Figure 6.9 ■ (a) Conformations of a four-carbon fragment of a polypropylene chain; (b) heptad of polypropylene chain: observed methyl is marked by *.

the Suter–Flory (1975) RIS model for PP and the resultant probabilities of finding CH_3 in a *gauche* arrangement with its γ-substituents (C_α) are multiplied by the shielding produced by this arrangement ($\gamma_{CH_3, C_\alpha} = -5$ ppm), we obtain the predicted methyl ^{13}C chemical shifts presented in the form of a stick spectrum at the bottom of Figure 6.10.

Because the γ-*gauche*-effect method of calculating ^{13}C chemical shifts only leads to the prediction of stereosequence-dependent relative chemical shifts, we are free in the comparison with observed spectra to translate the calculated shifts to obtain the best agreement with the observed $\delta^{13}C$'s. This has been done in Figure 6.10, where the agreement between the observed and calculated methyl $\delta^{13}C$'s has been used to make the stereosequence assignments indicated there. The γ-*gauche*-effect method of assigning resonances in the methyl region of the ^{13}C NMR spectrum of *a*-PP to heptad stereosequences has been achieved without recourse to the study of PP model compounds or stereoregular PPs and without assuming a particular statistical model to describe the frequencies of stereosequences produced during polymerization.

By achieving agreement between the observed ^{13}C chemical shifts and those predicted by the γ-*gauche*-effect method we have not only determined the microstructure (stereosequence) of this polymer, but in addition we have stringently tested its conformational characteristics as embodied in the RIS

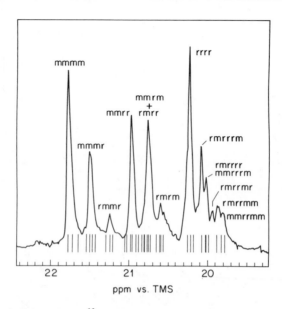

Figure 6.10 ■ The 90-MHz methyl ^{13}C spectrum of atactic polypropylene, observed in 1,2,4-tri-chlorobenzene at 100°C. [Reprinted with permission from Schilling and Tonelli (1980).]

model. Clearly then it is possible to use ^{13}C NMR spectroscopy to test or derive the conformational characteristics of vinyl polymers by comparison of observed spectra with the δ^{13}C's calculated via the γ-*gauche*-effect method. This approach has been pursued with success to test the local conformational characteristics of several vinyl polymers (Tonelli, 1978a, b; 1979; 1985; Tonelli and Schilling, 1981).

6.5. Establishing Vinyl Polymerization Mechanisms from Stereosequence Analysis

Having established the assignment of resonances observed in the methyl region of the ^{13}C NMR spectrum of *a*-PP to the appropriate heptad stereosequences, one might ask what use can be made of this detailed configurational information. Through an analysis of the intensities of the observed resonances we may determine if any simple statistical model, such as Bernoullian or Markovian statistics, can describe the polymerization of *a*-PP (see Section 6.2.4). In Figure 6.11 we compare the observed and simulated ^{13}C NMR spectra of the methyl region of *a*-PP. The simulated spectrum was obtained (Tonelli and Schilling, 1981) by assuming Lorentzian peaks of < 0.1 ppm width at half height for each of the 36 heptad chemical shifts calculated by the

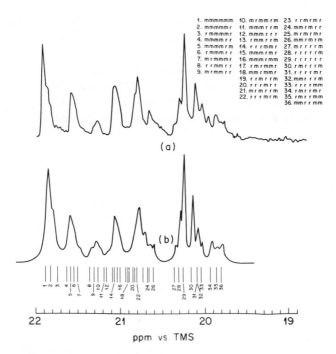

1. mmmmmm	10. m r mm r m	23. r r m r m r
2. mmmmm r	11. mmm r r m	24. mm r m r r
3. r mmmm r	12. mmm r r r	25. m r m r m r
4. mmmm r r	13. r mm r r m	26. mm r m r m
5. mmmm r m	14. r r r mm r	27. m r r r r m
6. r mmm r r	15. mmm r m r	28. r r r r r m
7. m r mmm r	16. mmm r mm	29. r r r r r r
8. r r mm r r	17. r m r mm r	30. r m r r r m
9. m r mm r r	18. mm r mm r	31. r r r r m r
	19. r r m r r m	32. mm r r r m
	20. r r m r r r	33. r r r r r m
	21. m r m r r m	34. r m r r m r
	22. r r r m r m	35. r m r r mm
		36. mm r r mm

(a)

(b)

22   21   20   19

ppm vs TMS

Figure 6.11 ■ (a) ^{13}C NMR spectrum at 90.52 MHz of the methyl carbon region in atactic PP in 20% w/v n-heptane solution at 67°C. (b) Simulated spectrum obtained from calculated chemical shifts, as represented by the line spectrum below, assuming Lorentzian peaks of < 0.1 ppm width at half height. [Reprinted with permission from Tonelli and Schilling (1981).]

γ-*gauche*-effect method. The relative intensities or heights of these heptad peaks were then adjusted to obtain the best simulation of the observed spectrum.

The comparison presented in Figure 6.11 makes apparent that we have been able to successfully simulate the methyl region of the ^{13}C NMR spectrum of a-PP based on our ability to calculate and assign all of the heptad stereosequence resonances. Thus, from this successful simulation we know how much of each heptad stereosequence is present in our a-PP sample. When we compare these heptad stereosequence frequencies with those predicted by the simple statistical models mentioned in Section 6.2.4, we are able to conclude (Schilling and Tonelli, 1980) that our a-PP sample cannot be described by any simple statistical polymerization model, such as Bernoullian or first-order Markovian.

It has been subsequently shown by Inoue et al. (1984) that a two-site model of the Ziegler–Natta polymerization of propylene (Zakharov et al., 1983) adequately describes the distribution of stereosequences observed in a-PP. At one site the monomer addition obeys Bernoullian statistics, and at the other site a predominance of monomer units are added in only one of the two possible $(0,1$ or $d, l)$ configurations.

As we can see, the γ-*gauche* prediction of ^{13}C NMR chemical shifts in vinyl polymers permits assignment of their ^{13}C NMR spectra, provides an opportunity to test or derive a RIS model description of their conformational characteristics, and may also permit a test of their polymerization statistics.

References

Bax, A. (1983). *J. Magn. Reson.* **53**, 517.

Bovey, F. A. (1969). *Polymer Conformation and Configuration*, Academic Press, New York.

Bovey, F. A. (1972). *High Resolution NMR of Macromolecules*, Academic Press, New York.

Bovey, F. A. (1982). *Chain Structure and Conformation of Macromolecules*, Academic Press, New York.

Bovey, F. A. and Mirau, P. A. (1988). *Accts. Chem. Res.* **21**, 37.

Bruch, M. D., Bovey, F. A., and Cais, R. E. (1984). *Macromolecules* **17**, 2547.

Cais, R. E. and Kometani, J. M. (1984). In *NMR and Macromolecules*, J. C. Randall, Ed., Symp. Ser. No. 247, Am. Chem. Soc., Washington, D.C., p. 153.

Carman, C. J. (1973). *Macromolecules* **6**, 725.

Chang, C., Muccio, D. D., and St. Pierre, T. (1985). *Macromolecules* **18**, 2334.

Dworak, A., Freeman, W. J., and Harwood, H. J. (1985). *Polymer J.* **17**, 351.

Frisch, H. L., Mallows, C. L., and Bovey, F. A. (1966). *J. Chem. Phys.* **45**, 1565.

Inoue, Y., Itabashi, Y., Chûjû, R., and Doi, Y. (1984). *Polymer* **25**, 1640.

Ishihara, N., Semimiya, T., Kuramoto, M., and Uoi, M. (1986). *Macromolecules* **19**, 2464.

Jeener, J. (1971). Presented at Ampère International Summer School, Basko Polje, Yugoslavia.

Price, F. P. (1962). *J. Chem. Phys.* **36**, 209.

Randall, J. C. (1977). *Polymer Sequence Determination*, Academic Press, New York.

Roberts, J. D. (1959). *Nuclear Magnetic Resonance: Applications to Organic Chemistry*, McGraw-Hill, New York, Chapter 3.

Schilling, F. C. and Tonelli, A. E. (1980). *Macromolecules* **13**, 270.

Shepherd, L., Chen, T. K., and Harwood, H. J. (1979). *Polym. Bull.* **1**, 445.

Stehling, F. C. and Knox, J. R. (1975). *Macromolecules* **8**, 595.

Suter, U. W. and Flory, P. J. (1975). *Macromolecules* **8**, 765.

Suter, U. W. and Neuenschwander, P. (1981). *Macromolecules* **14**, 528.

Tonelli, A. E. (1978a). *Macromolecules* **11**, 565.

Tonelli, A. E. (1978b). *Macromolecules* **11**, 634.

Tonelli, A. E. (1979). *Macromolecules* **12**, 255.

Tonelli, A. E. (1985). *Macromolecules* **18**, 1086.

Tonelli, A. E. and Schilling, F. C. (1981). *Accts. Chem. Res.* **14**, 233.

Tonelli, A. E., Schilling, F. C., and Cais, R. E. (1982). *Macromolecules* **15**, 849.

Weigert, F. (1971). *J. Org. Magn. Reson.* **3**, 373.

Wilson, C. W., III, and Santee, E. R., Jr. (1965). *J. Polym. Sci. Part C* **8**, 97.

Wüthrich, K. (1986). *NMR of Proteins and Nucleic Acids*, Wiley, New York.

Zakharov, V. A., Bukatov, G. P., and Yermakov, Y. I. (1983). *Adv. Polym. Sci.* **51**, 61.

Zambelli, A., Locatelli, P., Bajo, G., and Bovey, F. A. (1975). *Macromolecules* **8**, 687.

Microstructural Defects in Polymers

7.1. Introduction

During the polymerization process, as each monomer unit is added to the growing polymer chain end, it may be possible to enchain the monomer in more than a single structure. The stereosequence of monomer unit enchainment in vinyl polymers, as discussed in Chapter 6, is an example where each monomer may be added in two distinct ways to form m and r diads. If all monomer additions are of a single kind, all m or all r, then we obtain the stereoregular forms of the vinyl polymer, isotactic or syndiotactic. Atactic vinyl polymers possessing a distribution of m or r diad sequences may be considered microstructurally defective.

Unsymmetrically substituted vinyl monomers ($CHR{=}CH_2$) may in principle be added to the growing chain, or propagate, in either a head-to-tail (H–T) (a) or head-to-head : tail-to-tail (H–H : T–T) (b) mode:

$$
\text{(a)} \quad \cdots -CH_2-\underset{|}{\overset{|}{\underset{R}{C}H}}-CH_2-\underset{|}{\overset{|}{\underset{R}{C}H}}-CH_2-\underset{|}{\overset{|}{\underset{R}{C}H}}-CH_2-\underset{|}{\overset{|}{\underset{R}{C}H}}-\cdots
$$

$$
\text{(b)} \quad \cdots -CH_2-\underset{R}{C}H-\underset{R}{C}H-CH_2-CH_2-\underset{R}{C}H-\underset{R}{C}H-CH_2-\cdots
$$

In practice it is observed that the H–T enchainment of monomer units (a) is overwhelmingly preferred and the H–H : T–T enchainment (b) is generally relegated to an occasional inverted unit:

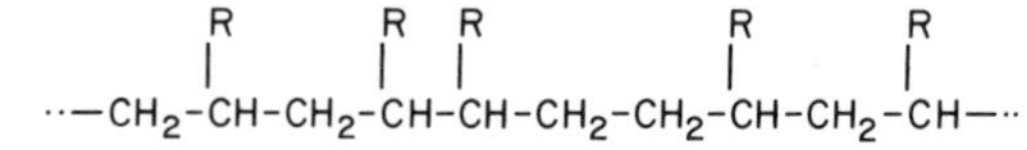

$$\cdots-CH_2-\underset{\underset{R}{|}}{CH}-CH_2-\underset{\underset{R}{|}}{CH}-\underset{\underset{R}{|}}{CH}-CH_2-CH_2-\underset{\underset{R}{|}}{CH}-CH_2-\underset{\underset{R}{|}}{CH}-\cdots$$

When discussing the stereosequences in the vinyl fluoropolymer PVF, it was mentioned that about 10% of the monomer additions are inverted during the commercial polymerization. The vinyl fluoropolymers are unusual in this regard (Tonelli et al., 1982), because most vinyl polymers are predominantly isoregic and contain many fewer inverted monomer units.

Another microstructural defect commonly introduced during polymerization is branching. If during the polymerization of ethylene ($CH_2{=}CH_2$), for example, the radical at the growing end of the polymer chain is transferred via hydrogen abstraction to an interior unit of the chain, then a branch is formed upon subsequent monomer additions:

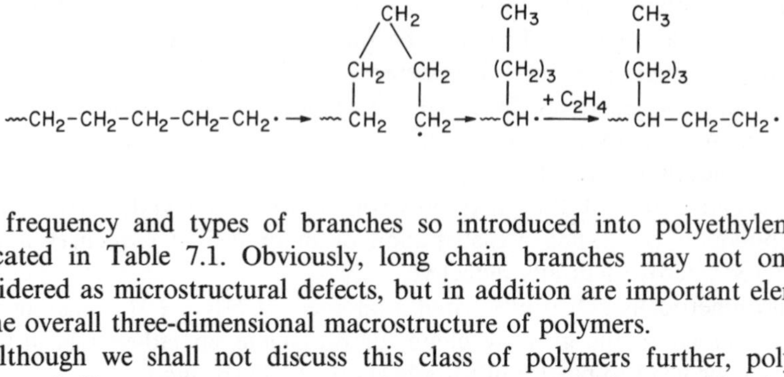

The frequency and types of branches so introduced into polyethylene are indicated in Table 7.1. Obviously, long chain branches may not only be considered as microstructural defects, but in addition are important elements of the overall three-dimensional macrostructure of polymers.

Although we shall not discuss this class of polymers further, polymerization of diene monomers can produce structures having combinations of

Table 7.1 ▪ Branching in High-Pressure Polyethylene[a]

Type of branch	Number of branches/1000 backbone carbons
$-CH_3$	0.0
$-CH_2CH_3$	1.0
$-CH_2CH_2CH_3$	0.0
$-CH_2CH_2CH_2CH_3$	9.6
$-CH_2CH_2CH_2CH_2CH_3$	3.6
Hexyl or longer	5.6
	19.8

[a] Bovey and Kwei (1979).

geometrical, stereochemical, and regiochemical isomerism. Butadiene, $CH_2=CH-CH=CH_2$, can be enchained in 1,4 or 1,2 fashion, with (a) *cis* (Z) or (b) *trans* (E) structures, and (c) isotactic or (d) syndiotactic stereosequences:

$$\begin{array}{cc}
\underset{\cdots-CH_2}{\overset{H}{\diagdown}}C=C\underset{CH_2-\cdots}{\overset{H}{\diagup}} & \underset{\cdots-CH_2}{\overset{H}{\diagdown}}C=C\underset{H}{\overset{CH_2-\cdots}{\diagup}} \\
\text{1,4-cis (Z)} & \text{1,4-trans (E)} \\
(a) & (b)
\end{array}$$

$$\begin{array}{cc}
\cdots-CH_2-\underset{\underset{CH_2}{\overset{\|}{CH}}}{CH}-CH_2-\underset{\underset{CH_2}{\overset{\|}{CH}}}{CH}-\cdots & \cdots-CH_2-\underset{\underset{CH_2}{\overset{\|}{CH}}}{CH}-CH_2-\underset{\overset{\underset{CH}{\overset{\|}{CH_2}}}{|}}{CH}-\cdots \\
\text{1,2-ISOTACTIC} & \text{1,2-SYNDIOTACTIC} \\
(c) & (d)
\end{array}$$

Substituted dienes can result in even more complicated structures combining all three forms of isomerism (Bovey, 1982).

In this chapter we will focus on the NMR study of monomer inversion and the various microstructures, or regiosequences which result. Two polymers, poly(vinylidene fluoride) (PVF_2) and poly(propylene oxide) (PPO), will serve as subjects to illustrate the techniques used to unravel the regiosequences of polymers containing inverted monomer units. Among these techniques, the γ-*gauche*-effect method of predicting ^{13}C and ^{19}F chemical shifts and their use in determining polymer regiosequences will be emphasized.

7.2. Determining the Regiosequence of PVF_2

Poly(vinylidene fluoride) (PVF_2), $-(CF_2-CH_2)-$, as normally obtained by free-radical polymerization, is not completely regioregular. Occasional inverted or reversed monomer units are incorporated during normal polymerization, resulting in the creation of between 3.5 and 6% defects or regioirregularities (Görlitz et al., 1973). Cais and Kometani (1983, 1985) have synthesized completely regioregular PVF_2 with no inverted monomer units through reduction of poly(1,1-dichloro-2,2-difluoroethylene) and PVF_2 samples containing as many as 23% defects by copolymerization of vinylidene fluoride with either 1-chloro-2,2-difluoroethylene or 1-bromo-2,2-difluoroethylene followed by reductive dehalogenation with tri-*n*-butyltin hydride. As we will subsequently

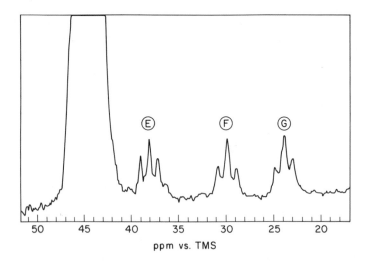

Figure 7.1 ■ The 25-MHz spectrum of the CH_2 carbons of poly(vinylidene fluoride) observed at 90°C using a 30% (w/v) solution in ethylene carbonate at 90°C. The very large resonance centered at about 45 ppm is that of the normal head-to-tail sequences. The "defect" resonances E, F, and G are identified in the text. [Reprinted with permission from Bovey et al. (1977).]

discuss, 2D NMR (^{19}F COSY) studies of the regiosequences in PVF_2 become feasible for samples with high defect contents.

7.2.1. ^{13}C NMR

Spin–spin coupling between directly bonded ^{13}C and ^{19}F nuclei generally obliterates the detailed structure of the dispersion in ^{13}C NMR chemical shifts produced by different fluoropolymer microstructures. Only in the methylene carbon regions of the usual ^{1}H-decoupled ^{13}C NMR spectra of PVF_2 and PVF can we begin to separate the effects of ^{19}F coupling and microstructure on the ^{13}C chemical-shift dispersion. The usual ^{1}H-decoupled ^{13}C NMR spectrum of PVF_2 is presented in Figure 7.1 (Bovey et al., 1977). Note the defect resonances E, F, and G:

$$\overset{\text{E}}{\cdots - CH_2CF_2CH_2CF_2}\overset{}{CH_2CF_2CF_2}\overset{\text{G}\text{F}}{CH_2CH_2CF_2}CH_2CF_2 -\cdots$$

corresponding to inverted monomer units. (Their assignment will be established below.)

Each of the methylene carbon resonances appears as a multiplet; the H–T and E resonances are pentuplets because of four neighboring fluorines, while F and G are triplets because they have only two neighboring fluorines. Schilling (1982) has developed a triple-resonance scheme to record the ^{13}C NMR spectra of fluoropolymers that are free of both $^{13}C-^1H$ and $^{13}C-^{19}F$ J-couplings. This is achieved by the simultaneous broad-band decoupling of both proton and fluorine nuclei while observing the carbon nuclei, and yields ^{13}C NMR spectra in which all carbon nuclei are fully decoupled and appear as single resonances. Figure 7.2 presents the completely decoupled ^{13}C NMR spectrum of PVF_2 (Tonelli et al., 1981) obtained by Schilling's (1982) triple-resonance technique. Note the absence of the usual multiplet pattern of ^{19}F-coupled ^{13}C resonances. Even the CF_2 carbon resonances appear as single peaks, though they possess two directly bonded fluorines. We can now begin to analyze the regiosequence of PVF_2 based on its ^{13}C NMR spectrum free of ^1H and ^{19}F J-coupling, as shown in Figure 7.2.

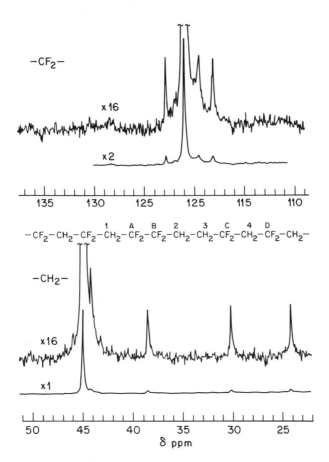

Figure 7.2 ■ The 22.5-MHz ^{13}C NMR triple-resonance spectrum of PVF_2. [Reprinted with permission from Tonelli et al. (1981).]

The absence of asymmetric centers in PVF$_2$ simplifies the analysis of its regiosequence defect structures. Occasional monomer inversions produce head-to-head : tail-to-tail (H–H : T–T) defects which are manifested as four resonances -0.8, -8, -15, and -21 ppm upfield from the regioregular head-to-tail (H–T) resonance in the methylene carbon region and by three resonances at 1.8, -1.5, and -2.9 ppm relative to the H–T CF peak (see Figure 7.2). Based on the PVF$_2$ fragment shown below,

$$- \, CF_2 \overset{a}{-} CH_2 \overset{b}{-} CF_2 \overset{c}{-} \overset{1}{CH_2} \overset{d}{-} \overset{A}{CF_2} \overset{e}{-} \overset{B}{CF_2} \overset{f}{-} \overset{2}{CH_2} \overset{g}{-} \overset{3}{CH_2} \overset{h}{-} \overset{C}{CF_2} \overset{i}{-} \overset{4}{CH_2} \overset{j}{-} \overset{D}{CF_2} \overset{k}{-} CH_2 \, -$$

we may write (Tonelli et al., 1981) the following expressions for the chemical shifts of the H–T methylene carbons, δ_{CH_2}, and those in the vicinity of H–H : T–T defects in terms of β and γ substituent effects and the bond rotation probabilities P which determine the frequencies of γ-*gauche* effects:

$$\delta_{CH_2}^{H-T} = 2(1 - P_t)\gamma_{CH_2,CF_2}, \tag{7.1}$$

$$\delta_{CH_2}^{1} = (1 - P_{b,t})\gamma_{CH_2,CF_2} + (1 - P_{e,t})\gamma_{CH_2,CH_2} + (1 + P_{e,t})\gamma_{CH_2,F}, \tag{7.2}$$

$$\delta_{CH_2}^{2} = (2 - P_{e,t} - P_{h,t})\gamma_{CH_2,CH_2} + (2 + P_{e,t} + P_{h,t})\gamma_{CH_2F} - \beta_{CH_2,F_2}, \tag{7.3}$$

$$\delta_{CH_2}^{3} = (2 - P_{f,t} - P_{i,t})\gamma_{CH_2,CF_2} + (1 + P_{f,t})\gamma_{CH_2,F} - \beta_{CH_2,F_2}, \tag{7.4}$$

$$\delta_{CH_2}^{4} = (1 - P_{h,t})\gamma_{CH_2,CH_2} + (1 - P_{k,t})\gamma_{CH_2,CF_2}. \tag{7.5}$$

As an example, let us derive the expression (7.4) for the ^{13}C chemical shift of the T–T methylene carbon $\overset{3}{C}H_2$. Newman projections about the bonds f and i in the PVF$_2$ fragment shown above are presented in Figure 7.3. Rotations about these bonds govern the γ-*gauche* interactions of $\overset{3}{C}H_2$. When bonds f and i are in the *gauche* rotational states ($g^\pm$), $\overset{3}{C}H_2$ is γ-*gauche* to $\overset{A}{C}F_2$ and $\overset{D}{C}F_2$. Hence the term $(2 - P_{f,t} - P_{i,t}) \times \gamma_{CH_2,CF_2}$ in Eq. 7.4. In addition, $\overset{3}{C}H_2$ is γ-*gauche* to F when $\phi_f = g^\pm$ and γ-gauche to two F's when $\phi_f = t$. Thus, the term $(1 + P_{f,t}) \times \gamma_{CH_2,F}$ accounts for all the fluorine shielding at $\overset{3}{C}H_2$. Compared to H–T methylene carbons, the T–T CH$_2$'s (2 and 3) have two fewer β-fluorine substituents, and this results in the term $-\beta_{CH_2,F_2}$ in Eq. 7.4.

Bond rotation probabilities P_t, $P_{b,t}$, $P_{e,t}$, $P_{f,t}$, $P_{h,t}$, $P_{i,t}$, and $P_{k,t}$ were obtained from the RIS conformational model developed for H–T and H–H : T–T PVF$_2$ (Toneli, 1976). Because $P_t(\text{H–T}) = P_{k,t}(\text{H–H : T–T})$ (see Figure 7.3), the difference between the ^{13}C chemical shifts of H–H : T–T methylene carbon 4 and H–T methylene carbons reduces to $\delta_{CH_2}^{4} - \delta_{CH_2}^{H-T}$
$= (1 - P_{h,t}) \times \gamma_{CH_2,CH_2} - (1 - P_t) \times \gamma_{CH_2,CF_2} = 0.348\gamma_{CH_2,CH_2} - 0.48\gamma_{CH_2,CF_2}$.
As we have seen from the ^{13}C NMR study of hydrocarbon polymers (see

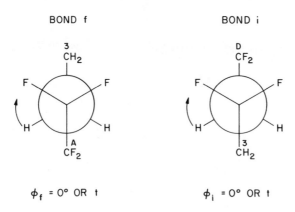

Figure 7.3 ■ Newman projections about bonds f and i in a PVF$_2$ fragment containing an inverted monomer unit.

Chapter 4), $\gamma_{CH_2, CH_2} = -5.3$ ppm, leading to $\delta_{CH_2}^{4} - \delta_{CH_2}^{H-T} = -1.85 - 0.48\gamma_{CH_2, CF_2}$. Now $\delta_{CH_2}^{4} - \delta_{CH_2}^{H-T}$ must equal -0.8 ppm, because if $\delta_{CH_2}^{4}$ were assigned to the defect resonances -7, -15, or -21 ppm upfield from $\delta_{CH_2}^{H-T}$ (see Figure 7.2), then γ_{CH_2, CF_2} would have to be -11, -27, or -40 ppm, respectively. Values this large (-11 to -40 ppm) for the shielding produced at CH$_2$ by a γ-*gauche* CF$_2$ group (γ_{CH_2, CF_2}) are clearly unreasonable in view of our experience (Tonelli et al., 1979; Tonelli and Schilling, 1981) with vinyl chloride oligomers, homopolymer, and copolymers. Thus, $\delta_{CH_2}^{4} - \delta_{CH_2}^{H-T} = -0.8$ ppm $= -1.85 - 0.48\gamma_{CH_2, CF_2}$, which leads to $\gamma_{CH_2, CF_2} = -2.2$ ppm.

H–H : T–T methylenes 2 and 3 must be farthest upfield from H–T CH$_2$, because they possess two fewer deshielding β-fluorine substituents. By elimination $\delta_{CH_2}^{1} - \delta_{CH_2}^{H-T} = -7$ ppm, which results in $\gamma_{CH_2, F} = -3.8$ ppm. Because $\overset{2}{C}H_2$ has twice as many shielding γ-fluorines as $\overset{3}{C}H_2$, we are led to conclude that $\delta_{CH_2}^{2} - \delta_{CH_2}^{H-T} = -21$ ppm and $\delta_{CH_2}^{3} - \delta_{CH_2}^{H-T} = -15$ ppm. These expressions result in $\beta_{CH_2, F_2} = +8$ ppm. The β and γ effects ($\beta_{CH_2, F_2} = +8$ ppm, $\gamma_{CH_2, CF_2} = -2.2$ ppm, and $\gamma_{CH_2, F} = -3.8$ ppm) derived by comparison of observed H–T and H–H : T–T methylene carbon chemical shifts result in the line spectrum drawn below the observed methylene ^{13}C NMR spectrum of PVF$_2$ presented in Figure 7.4.

An analysis similar to that described above for the CH$_2$ carbons, when applied to the ^{13}C chemical shifts of the quaternary CF$_2$ carbons, leads to the calculated line spectrum drawn below the observed CF$_2$ region of the PVF$_2$ spectrum in Figure 7.4. The following substituent effects are deduced from the

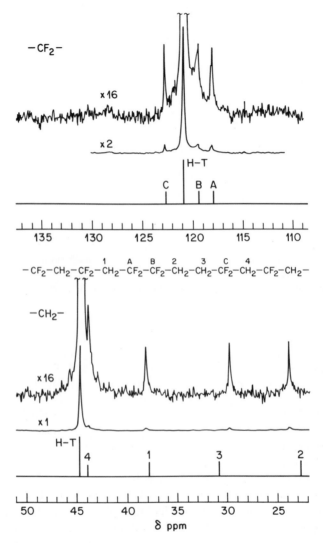

Figure 7.4 ■ ^{13}C NMR spectrum and calculated ^{13}C chemical shifts for PVF_2. [Reprinted with permission from Tonelli et al. (1981).]

comparison of observed and calculated CF_2 chemical shifts: $\gamma_{CF_2,C} = -2.1$ ppm, $\gamma_{CF_2,F} = -1.4$ ppm, and $\beta_{CF_2,F_2} = -5$ ppm. The γ-effects experienced by the CF_2 carbons are less than half as large as those governing the CH_2 carbon chemical shifts.

An even more striking difference exists between the effects produced by a pair of fluorine atoms in the β-position upon the chemical shifts of the CH_2 and CF_2 carbons in PVF_2: $\beta_{CH_2,F_2} = +8$ ppm and $\beta_{CF_2,F_2} = -5$ ppm. The

β-fluorine atoms deshield the CH_2 carbons, as is usually observed (Stothers, 1972) for β-substituents, but the CF_2 carbons are shielded by β-fluorine atoms. This dependence of the β-fluorine substituent effect on the degree of fluorine substitution at the observed carbon is also observed for chlorone substituents (Tonelli et al., 1981). Measurement of the ^{13}C NMR spectra of ethane and all nine of its chlorinated derivatives led to the following observations: $\beta^0_{1,2,3} = +12.2, +26.3,$ and $+40.0$ ppm; $\beta^1_{1,2,3} = +4.0, +10.4,$ and $+19.0$ ppm; $\beta^2_{1,2,3} = +1.3, +4.9,$ and $+10.6$ ppm; and $\beta^3_{1,2,3} = +1.0, +4.4,$ and $+9.9$ ppm, where for example, β^1_2 signifies the effect of two β-Cl substituents on the ^{13}C chemical shift of a carbon with a single α-Cl substituent.

To further test the validity of the derived shielding effects produced by a pair of fluorine atoms β on CF_2 carbons and the effects of carbon and fluorine atoms γ on CF_2, we calculated the ^{13}C chemical shift expected at the central carbons in perfluorooctane and compared it with the CF_2 chemical shifts in PVF_2. The β and γ effects derived for the CF_2 carbons in PVF_2 were used along with the conformational RIS model derived and tested for the perfluorinated alkanes and poly(tetrafluoroethylene) by Bates and Stockmayer (1968). The H–T CF_2 and H–H:T–T $\overset{\text{C}}{CF_2}$ carbons in PVF_2 do not have β-fluorine substituents (see Figure 7.4), while H–H:T–T CF_2 carbons A and B in PVF_2 and the central carbons in perfluorooctane (PFO) possess two and four β-fluorines, respectively. The following relative ^{13}C chemical shifts are calculated for the CF_2 carbons in H–T PVF_2, H–H:T–T (A, B, C), PVF_2, and PFO: $\delta^{PFO}_{CF_2} - \delta^A_{CF_2} = -6.6,$ $\delta^{PFO}_{CF_2} - \delta^B_{CF_2} = -8.1,$ $\delta^{PFO}_{CF_2} - \delta^{H-T}_{CF_2} = -9.5,$ and $\delta^{PFO}_{CF_2} - \delta^C_{CF_2} = -11.3$ ppm. These compare well with the ^{13}C chemical shifts observed here for the CF_2 carbons in PVF_2 and the central CF_2 carbons in neat PFO reported by Lyerla and VanderHart (1976), i.e., $\delta^{PFO}_{CF_2} - \delta^A_{CF_2} = -6.2,$ $\delta^{PFO}_{CF_2} - \delta^B_{CF_2} = -7.7,$ $\delta^{PFO}_{CF_2} - \delta^{H-T}_{CF_2} = -9.1,$ and $\delta^{PFO}_{CF_2} - \delta^C_{CF_2} = -11.0$ ppm. This comparison strengthens our confidence in the β- and γ-substituent effects derived for the CF_2 carbons in PVF_2 and permits an analysis of the ^{13}C NMR spectra of other fluoropolymers (Tonelli et al., 1981).

7.2.2. ^{19}F NMR

Having successfully analyzed the ^{13}C NMR spectrum of PVF_2 containing a small number of inverted units [3.2% H–H:T–T monomer addition as obtained by integration of defect (H–H:T–T) and normal (H–T) resonances (Tonelli et al., 1981)], we now attempt to assign and analyze the ^{19}F NMR spectrum of the same PVF_2 sample. The ^{19}F NMR spectrum of PVF_2 measured at 84.6 MHz is presented in Figure 7.5(a). Three small resonances appear 3.2, 22.0, and 24.0 ppm upfield from the main H–T fluorine resonance at 91.6 ppm (relative to $CFCl_3$) and are attributed to the fluorine nuclei belonging to H–H:T–T inverted units (Wilson, 1963; Wilson and Santee, 1965).

For the PVF$_2$ fragment below,

$$-CF_2 \overset{a}{-} CH_2 \overset{b}{-} \overset{1}{CF_2} \overset{c}{-} CH_2 \overset{d}{-} \overset{2}{CF_2} \overset{e}{-} \overset{3}{CF_2} \overset{f}{-} CH_2 \overset{g}{-} CH_2 \overset{h}{-} \overset{4}{CF_2} \overset{i}{-} CH_2 \overset{j}{-} CF_2 \overset{k}{-} CH_2 -$$

we may write expressions for the relative ^{13}F NMR chemical shifts (δ_F) of the H–T and H–H : T–T fluorines in terms of the γ-effects ($\gamma_{F,F}$ and $\gamma_{F,C}$) and the bond rotation probabilities (P) which determine the frequencies of their γ-*gauche* interactions:

$$\delta_F^{H-T} = (1 + P_t)\gamma_{F,C}, \tag{7.6}$$

$$\delta_F^2 = (1 + 0.5P_{t,d} + 0.5P_{t,e})\gamma_{F,C} + (1.5 - 0.5P_{t,e})\gamma_{F,F}, \tag{7.7}$$

$$\delta_F^3 = (1 + 0.5P_{t,e} + 0.5P_{t,f})\gamma_{F,C} + (1.5 - 0.5P_{t,e})\gamma_{F,F}, \tag{7.8}$$

$$\delta_F^4 = (1 + 0.5P_{t,h} + 0.5P_{t,i})\gamma_{F,C}. \tag{7.9}$$

Comparison of Eq. 7.6 and Eq. 7.9 reveals that δ_F^{H-T} and δ_F^4 are most similar. Thus, $\delta_F^4 - \delta_F^{H-T} = 3.2$ ppm, which leads directly to $\gamma_{F,C} = +30$ ppm (shielding). By elimination, $|\delta_F^2 - \delta_F^3| = 2$ ppm, which yields $\gamma_{F,F} = +15$ ppm. Substitution of $\gamma_{F,F} = 15$ ppm and $\gamma_{F,C} = 30$ ppm into Eqs. 7.6–9 leads to calculated δ_F's shown as sticks in (c) of Figure 7.5, and which compare well with the observed spectrum in (a) recorded at low field strength.

At 188 MHz four additional defect resonances (1, 5, 6, and 7) appear in the ^{13}F NMR spectrum of PVF$_2$ [see (b) of Figure 7.5]. Ferguson and Brame (1979) also observed these additional defect peaks and tentatively assigned them to defect structures drawn in (b) based on α, β, and γ substituent effects derived from the CF$_2$ resonances observed in various saturated, partially fluorinated linear alkanes. In addition to δ_F^{H-T}, δ_F^2, δ_F^3, and δ_F^4, ^{19}F chemical shifts were also calculated for the defect CF$_2$ fluorines 1, 5, 6, and 7. ^{19}F γ-effects (γ_{F,CH_2}, γ_{F,CF_2}, and $\gamma_{F,F}$) were least-squares fitted to produce the best agreement between observed and calculated δ_F's [see (b) and (c) of Figure 7.5]. Best agreement was achieved for $\gamma_{F,CH_2} = \gamma_{F,CF_2} = \gamma_{F,C} = 25-30$ ppm and $\gamma_{F,F} = 15$ ppm, confirming the assignments proposed by Ferguson and Brame (1979) and the γ-effects derived from the defect resonances 2, 3, and 4 observed at lower magnetic field strength.

Measurement of the intensities of the defect resonances and comparison with the total intensity of all observable resonances yields an estimate of 3.4% defect H–H : T–T addition in this sample of PVF$_2$. This compares well with the 3.2% defect estimate described earlier using the ^{13}C NMR analysis.

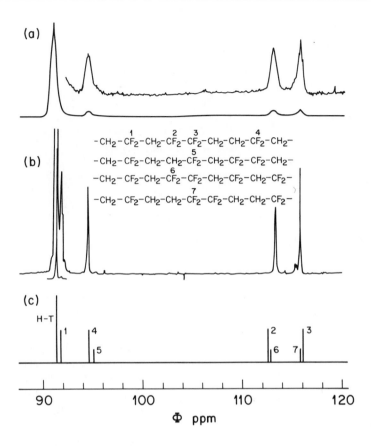

Figure 7.5 ■ Observed and calculated ^{19}F NMR spectra of PVF$_2$: (a) measured at 84.6 MHz; (b) measured at 188.2 MHz; (c) calculated. Vertical expansion in (a) is ×8, in (b) ×40. [Reprinted with permission from Tonelli et al. (1982).]

7.2.3. 2D ^{19}F NMR

2D ^{19}F NMR J-correlated (COSY) experiments are very difficult to perform on commercial PVF$_2$ samples containing from 3 to 6% reversed units (Ferguson and Ovenall, 1984; Cais and Kometani, 1985). The aregic H–H : T–T resonances are weak, so that the 2D cross peaks are easily confused with artifacts generated by the intense ridges (Bax, 1982) radiating from the main H–T isoregic peak. However, Cais and Kometani (1985) have synthesized highly aregic PVF$_2$ (up to 23% defects) and have been able to verify in an absolute manner the main regiosequence assignments in their ^{19}F NMR spectra through the use of J-correlated (2D COSY) ^{19}F NMR spectroscopy.

Scheme I

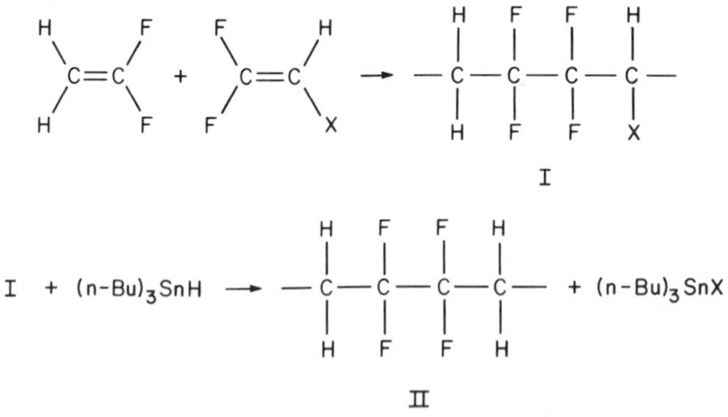

$$X = Cl \text{ or } Br$$

Cais and Kometani (1985) utilized the synthetic procedure outlined in Scheme I to obtain highly aregic PVF_2. They found the chlorinated and brominated monomers to be attacked exclusively at the CF_2 carbon by the growing $-CH_2-CF_2 \cdot$ radical, so that after reductive dehalogenation with tri-n-butyltin hydride they become inverted VF_2 units in the polymer II, which is a regioisomer of PVF_2. They were able to control the level of reversed H–H : T–T units introduced by varying the feed ratio in the initial copolymerization mixture. the 470-MHz ^{19}F NMR spectra of three of their aregic PVF_2 samples are presented in Figure 7.6. The assignment of resonances to PVF_2 regiosequences is the same as proposed by Ferguson and Brame (1979) and Tonelli et al. (1982).

The 2D J-coupled (2D COSY) ^{19}F NMR spectrum of aregic PVF with 18% reversed units is displayed in Figure 7.7. The off-diagonal or cross peaks manifest a connectivity between fluorine pairs which have a nonzero scalar coupling transmitted through either three (vicinal) or four bonds (Bruch et al., 1984). Longer-range couplings are negligible and can be ignored. Since geminal fluorines in PVF_2 are magnetically equivalent, the large two-bond coupling expected is not observed (Cais and Kometani, 1984). The observed cross peaks can be rationalized on the basis of heptad regiosequences as assigned in the one-dimensional ^{19}F NMR spectra shown in Figure 7.6.

Allowing three- and four-bond homonuclear ^{19}F coupling, it is easy to examine the heptad regiosequences and establish their connectivity map. The necessary connectivities are A–B, A–C, B–C, B–E, B–F, D–E, D–F (four-bond), and E–H (three-bond). Implicit in this connectivity scheme is the correctness of the A–H regiosequence assignments shown in Figure 7.6, proposed by Ferguson and Brame (1979) and calculated empirically by Tonelli et al. (1982), which can now be rigorously tested by the 2D COSY experiment.

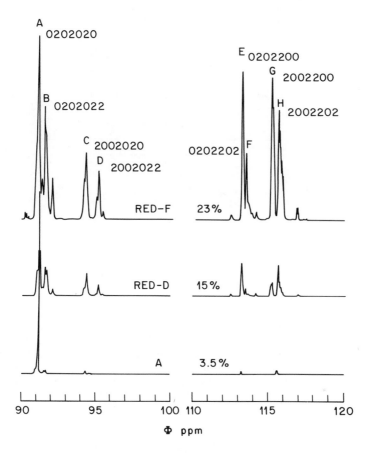

Figure 7.6 ■ Proton-decoupled 470-MHz fluorine-19 NMR spectra of aregic PVF$_2$ samples A, red-D, and red-F having 3.5, 15, and 23% defects, respectively. The samples were observed at 25°C as 10% solutions in dimethylformamide-d_7 on a JEOL GX-500 spectrometer. The sweep width was 20 kHz with 132K points, and 800 transients were accumulated with a pulse delay of 10 sec using 90° (8-μsec) pulses. The eight distinct regiosequence heptads (A–H) are assigned according to Cais and Sloane (1983) to the carbon sequences as shown (O = CH$_2$, 2 = CF$_2$). Most heptad peaks have quartet fine structure indicating a higher-order sensitivity to 11-carbon sequences (monomer sequence hexads). The three spectra are vertically scaled so that the isoregic peak A has the same intensity in each. [Reprinted with permission from Cais and Kometani (1985).]

All expected connectivities are observed in Figure 7.7 except for B–F, which can only be established within the regiosequence 0202022020 (0 = CH$_2$, 2 = CF$_2$) having two adjacent reversed units. Because — CF$_2$—CHX— units are never adjacent in the precursor copolymer (Cais and Kometani, 1985), the likelihood of two consecutive monomer reversals is low, so the cross peak is weak and not observed in the 2D COSY experiment.

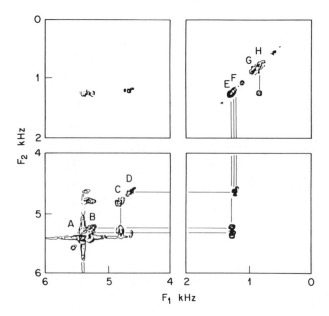

Figure 7.7 ▪ Absolute-value contour plot of the 188-MHz fluorine-19 2D *J*-correlated spectrum of aregic PVF$_2$ with 18% defects. The polymer was observed as a 10% solution in dimethylformamide-d_7 at 30°C on a Varian XL-200 with broad-band proton decoupling. A total of 64 transients was accumulated for each of 256 spectra with 1024 points covering 7000 Hz in both dimensions. The central region (2000 × 2000 Hz) contains no signals and has been removed for clarity. [Reprinted with permission from Cais and Kometani (1985).]

Assignments of peaks A and G (Cais and Kometani, 1983, 1985) are obtained from the model isoregic and syndioregic (alternating ethylene–tetrafluoroethylene copolymer) polymers. Having established these two assignments and using the connectivities observed in Figure 7.7, Cais and Kometani (1985) were able to uniquely establish the assignments of peaks B, C, D, E, F, and H, as also indicated there. These are in complete agreement with previous work (Ferguson and Brame, 1979; Tonelli et al., 1982; Cais and Sloane, 1983).

7.3. Regiosequence Defects in PPO

Propylene oxide

$$(CH_3\overset{\displaystyle \lceil O \rceil}{CH}CH_2)$$

exists in both *R* and *S* optical forms due to its asymmetric methine carbon. If during polymerization only one of the C–O bonds in the cyclic monomer is cleaved, then it is possible to generate four different stereochemical triads in the regioregular head-to-tail (H–T) PPO polymer. These H–T triads are presented in planar zigzag projection:

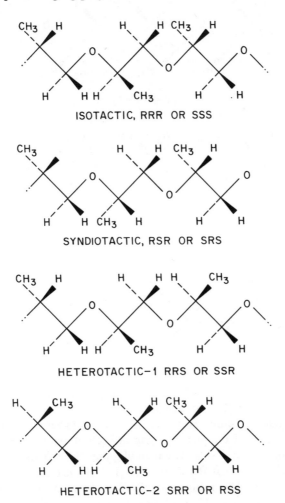

ISOTACTIC, RRR OR SSS

SYNDIOTACTIC, RSR OR SRS

HETEROTACTIC–1 RRS OR SSR

HETEROTACTIC–2 SRR OR RSS

If, however, during the ring-opening polymerization (Price and Osgan, 1956; Price et al., 1967) both C–O bonds in propylene oxide are subject to cleavage, then, in addition to the H–T PPO triads above, three additional structural triads or regiosequences are possible for PPO. These are illustrated here for the all-*R* regioisomers, where H–T, H–H, T–T, and T–H refer to the directions of neighboring monomers, and where H is the methine end and T is the methylene end of the monomer unit. Each of these regioirregular triads, with H–H and T–T additions, can be further subdivided on stereochemical grounds as was done above for the regioregular H–T triads. Thus, when both regiosequence and stereosequence are considered 16 unique structural triads can potentially exist in PPO.

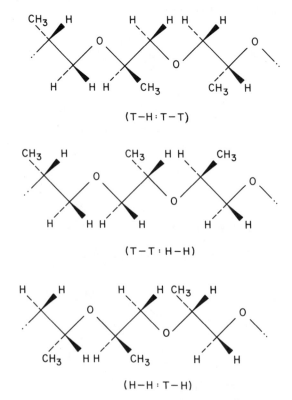

(T–H : T–T)

(T–T : H–H)

(H–H : T–H)

It is worth mentioning that independent of regiosequence (H–T, H–H, T–T), an *m*-diad consists of RR or SS neighboring units, while an *r*-diad consists of RS or SR neighboring units. However, the methyl groups in a H–T *m*-diad are on opposite sides of the planar zigzag projection, while in H–H and T–T diads they are on the same side. The methyl groups in *r*-diads are on the same side of the backbone when the diad is H–T, but on opposite sides in both H–H and T–T diads. This results directly from the number of bonds separating asymmetric centers in H–T (3 bonds) and in H–H, T–T (2, 4 bonds) regiosequences.

Because the PPO repeat unit contains three protons (two methylene and one methine) whose resonances overlap extensively, it has not been possible to use ^{1}H NMR spectroscopy (Ramey and Field, 1964; Tani et al., 1968; Hirano et al., 1972; Oguni et al., 1973a, b; Bruch et al., 1985), even at 500 MHz (Bruch et al., 1985), to determine the microstructure of PPO. Deuteration at the methine carbon simplifies the ^{1}H NMR spectra of PPO (Tani et al., 1968; Hirano et al., 1972; Oguni et al., 1973a, b), and 2D ^{1}H NMR spectroscopy (Bruch et al., 1985) also leads to greater separation of the overlapping proton resonances. However, even the application of these specialized synthetic and spectroscopic techniques has not been completely successful in establishing the microstructures present in PPO.

As we have seen, ^{13}C NMR generally offers the potential for greater spectroscopic resolution than ^{1}H NMR and might be expected to be better suited for the analysis of PPO microstructure (Schaefer, 1969; Oguni et al., 1972, 1979; Lapeyre et al., 1973; Uryu et al., 1973). This expectation is realized for regioregular (all H–T) PPO, where CH and CH_2 carbon resonances are separated by about 2 ppm, and permits the unambiguous assignment (Oguni et al., 1979) of PPO stereosequences. However, as we shall demonstrate here, the methine and methylene carbon resonances in regioirregular (H–T, H–H, T–T) PPO do overlap (Schilling and Tonelli, 1986).

The methine and methylene carbons in PPO have the same number and types of α and β substituents (CH $\rightarrow$ 2α-C, 1α-O, 1β-C, and 1β-O; CH_2 $\rightarrow$ 1α-C, 1α-O, 2β-C, and 1β-O) independent of whether or not they are part of H–T, H–H, or T–T units (see triad representations above). Because the deshielding of a carbon nucleus produced by α- and β-carbon substituents is virtually identical (about $+9$ ppm; see Section 4.2.2), the relative ^{13}C chemical shifts of both the CH and CH_2 carbons in PPO should depend solely on their γ-*gauche* interactions. In regioregular PPO the H–T methine carbons have two γ-substituents (2CH) and the methylene carbons three γ-substituents ($2CH_2, 1CH_3$). We therefore expect, as is observed (Oguni et al., 1979), that the methylene carbons resonate upfield (about 2 ppm) from the methine carbons. In regioirregular PPO the H–H methine carbons have three γ-substituents ($2CH_2$ and $1CH_3$ or 1CH, $1CH_2$, and $1CH_3$), as do the H–T methylene carbons, and the T–T methylene carbons have two γ-substituents (2CH or CH and CH_2), like the H–T methine carbons.

We therefore expect the H–H methine and H–T methylene carbon resonances and the T–T methylene and H–T methine resonances to overlap, based on their having the same number and types of α, β, and γ substituents. In earlier studies of PPO using ^{13}C NMR (Oguni et al., 1979), confusion developed in assigning resonances to carbons of the H–H : T–T defects that result from the catalyst occasionally cleaving the CH–O linkage instead of the CH_2—O bond when opening the propylene oxide ring. The possibility of mixing of the methine and methylene resonances was not considered. Additionally, spectral analysis of lower-molecular-weight samples must take into account the contributions of carbon nuclei in the chain-end structures.

Our approach (Schilling and Tonelli, 1986) in analyzing the ^{13}C NMR spectra of PPO was first to define the type of carbon represented by each resonance, i.e., methine, methylene, or methyl. Application of the DEPT and INEPT pulse-editing techniques [see Section 3.3 and Derome (1987)] permitted establishment of this objective. Second, by analysis of PPO samples differing in molecular weight we were able to assign those resonances belonging to end-group carbons. The third and final step in the analysis was to assign the H–H and T–T defect resonances using the γ-*gauche*-effect calculation of δ [13]C's described below.

The carbon nuclei in PPO are shielded by carbon and oxygen γ-substituents. From ^{13}C NMR studies of alkanes and their oxygenated derivatives [see Section 4.2.3 and Stothers (1972)], $\gamma_{C,C} = -4$ to -5 ppm and $\gamma_{C,O} = -6$ to -8 ppm seem likely for the shieldings produced by C and O γ-substituents when in a *gauche* arrangement with the carbon nuclei in PPO.

Table 7.2 ▪ Calculated ^{13}C NMR Chemical Shifts for Poly(propylene oxide) at 23°C

$$
\begin{array}{ccccc}
1 & 2 & 3 & 4 & 5 \\
CH_3\ \underline{a} & CH_3\ \underline{b}\ \ CH_3 & \underline{c} & CH_3\ \underline{d} & CH_3 \\
| & |\ \ \ \ | & & | & | \\
\end{array}
$$

$$-CH_2-CH-O-CH_2-CH-O-CH-CH_2-O-CH_2-CH-O-CH_2-CH-O-$$

$$
\begin{array}{cccccccccc}
1 & 1 & 2 & 2 & 3 & 3 & 4 & 4 & 5 & 5
\end{array}
$$

Carbon	Diad[a]				Chem. shift, ppm[b]
	a	*b*	*c*	*d*	
CH$_3$ 1	m	—	—	—	0.00
2	r	r	—	—	+0.45
2	m	r	—	—	+0.49
2	r	m	—	—	+0.75
2	m	m	—	—	+0.79
3	—	r	—	—	+0.53
3	—	m	—	—	+0.82
4	—	—	—	m	+0.02
4	—	—	—	r	+0.04
5	—	—	—	m	0.00
CH$_2$ 1	m	—	—	—	0.00
2	m	m	—	—	−0.15
2	r	m	—	—	−0.19
2	r	r	—	—	+0.20
2	m	r	—	—	+0.25
3	—	m	—	—	+4.40
3	—	r	—	—	+4.73
4	—	—	—	r	+4.75
4	—	—	—	r	+4.78
5	—	—	—	m	0.00
CH 1	m	—	—	—	0.00
2	m	m	—	—	−4.49
2	r	m	—	—	−4.45
2	m	r	—	—	−4.49
2	r	r	—	—	−4.45
3	—	m	—	—	−4.48
3	—	r	—	—	−4.48
4	—	—	—	r	−0.25
4	—	—	—	m	−0.27
5	—	—	—	m	0.00

[a] The dash (—) indicates either *m* or *r* diad placement.
[b] The + and − indicate downfield and upfield shifts, respectively, relative to the position of the 1 or 5 (H–T) carbons.

The numbers of such γ-*gauche* arrangements were determined from the bond conformation probabilities calculated for PPO with the RIS model developed by Abe et al. (1979). This conformational description developed for regioregular (H–T) PPO was modified so as to permit the calculation of bond probabilities in the H–H and T–T portions of PPO as well. Effects of both regiosequence and stereosequence were explicitly considered when calculating relative ^{13}C NMR chemical shifts in PPO via the γ-*gauche*-effect method. The results of these calculations for the carbon nuclei in the H–H and T–T defect structures in PPO are presented in Table 7.2. Note the significant differences between the δ^{13}C's predicted for the regular H–T and defect H–H and T–T carbons.

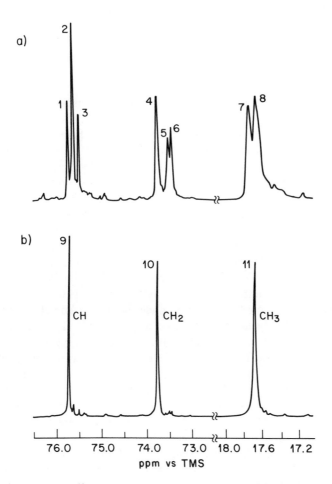

Figure 7.8 ■ 50.31-MHz ^{13}C NMR spectra of (a) atactic PPO 4000 and (b) isotactic PPO, observed at 23°C in C_6D_6. See Table 7.3. [Reprinted with permission from Schilling and Tonelli (1986).]

Table 7.3 ▪ ^{13}C NMR Chemical Shifts and Relaxation Data for Head-to-Tail Carbons in Poly(propylene oxide) at 23°C^a

Resonance	Chem. shift, ppm	T_1, sec	Carbon type	Stereosequence
1	75.75	0.78	CH	*mm*
2	75.64	0.80	CH	*mr + rm*
3	75.50	0.81	CH	*rr*
4	73.78	0.51	CH$_2$	*m*
5	73.54	0.50	CH$_2$	*r*
6	73.47	0.50	CH$_2$	*r*
7	17.79	1.03	CH$_3$	*rm, mr, rr*
8	17.71	1.03	CH$_3$	*mm, rm, mr, rr*
9	75.73		CH	*mm*
10	73.77		CH$_2$	*m*
11	17.72		CH$_3$	*mm*

^{a}Figure 7.8.

^{13}C NMR spectra of PPO 4000 ($M_n = 4000$) and isotactic PPO ($M_v = 14,500$) are presented in Figure 7.8. All three carbon types display chemical-shift sensitivity to the stereochemistry of the polymer chain. The assignments (see Table 7.3) of the regioregular H–T portion of PPO are made by comparison of the two spectra, and agree with earlier work (Oguni et al., 1972, 1979). In contrast to ^{13}C NMR observations for most vinyl polymers, the observed sensitivity of the PPO carbon chemical shifts to stereochemistry is very small. The total spread of δ ^{13}C's is only 0.12, 0.20, and 0.25 ppm for the methyl, methine, and methylene carbons respectively. This can be contrasted to polypropylene (Schilling and Tonelli, 1980), where the range of chemical shifts due to stereosequences is 2.0, 0.5, and 2.0 ppm for the same carbon types. The reduced sensitivity in PPO reflects the presence of three bonds between chiral centers in contrast to the two bonds in vinyl polymers. The limited chemical-shift sensitivity is predicted by the RIS model for PPO (Abe et al., 1979). On the basis of γ-*gauche* shielding interactions a spread of H–T chemical shifts of about 0.05 ppm is predicted for each of the three carbon types.

The DEPT technique (Turner, 1984; Derome, 1987) permits spectral editing in such a manner that spectra containing only a specific carbon type can be produced. In Figure 7.9 we show results of DEPT measurements on atactic PPO 4000 for the methine and methylene carbons only. At the high vertical gain used in acquiring these spectra the H–T resonances are off scale, and we are observing the resonances of the defect H–H and T–T structures as well as the chain-end carbons. In (a) all CH and CH$_2$ resonances are observed, while in (b) and (c) only the CH and CH$_2$ resonances, respectively, are observed. The most striking feature of these DEPT editing spectra is that there are clearly methine carbon resonances in the upfield region previously thought to

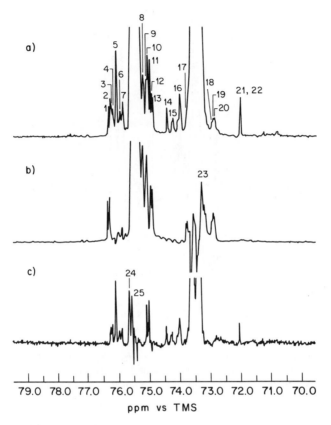

Figure 7.9 ■ Methine and methylene (a), methine only (b), and methylene only (c) 50.31-MHz ^{13}C NMR DEPT spectra of atactic PPO 4000 observed at 23°C in C_6D_6. [Reprinted with permission from Schilling and Tonelli (1986).]

contain exclusively methylene resonances, and there also are methylene resonances in the downfield portion of the spectra previously thought to contain only methine resonances. [These observations are confirmed by INEPT spectra (not shown) in which methylene resonances can be observed with negative intensity while methine signals appear as positive peaks.] Certain H–H : T–T and/or end-group resonances at about 73.5 and 75.6 ppm can only be observed in edited spectra, as they are completely obscured by the H–T peaks in a normal FT spectrum. The comparison of resonances in the three spectra of Figure 7.9 permits us to identify each resonance as to carbon type, methine or methylene.

To assign resonances produced by various end groups we compare the spectra observed for PPO 4000 (DP = 69) and PPO 1000 (DP = 17) as presented in Figure 7.10. Results of a DEPT measurement on PPO 1000 agree

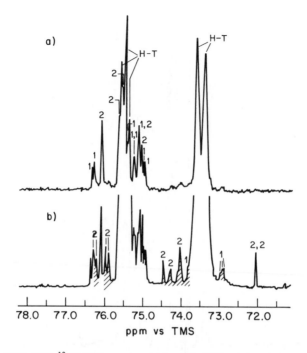

Figure 7.10 ▪ 50.31 MHz ^{13}C NMR spectra of (a) atactic PPO 1000 and (b) atactic PPO 4000 observed at 23°C in C_6D_6. (1 indicates methine and 2 indicates methylene.) The crosshatched resonances result from the H–H : T–T structure. [Reprinted with permission from Schilling and Tonelli (1986).]

with those of PPO 4000 in establishing the carbon type represented by each resonance. All of the visible resonances in Figure 7.10(a), other than the labeled H–T peaks, can be attributed to the end groups, because the number of such groups is about 3 times that of the H–H : T–T defects in the low-molecular-weight PPO 1000. All of the CH and CH_2 end-group resonances occur in the H–T methine region between 75.0 and 76.5 ppm.

DEPT spectra of PPO 1000 (not shown) indicate that no end-group CH resonances are hidden by the H–T methylene resonance at 73.5 ppm. Comparison of DEPT spectra permits the specific assignment of end-group methine (1) and methylene (2) resonances. Note in Figure 7.10(a) the end-group methylene resonances at about 75.6 ppm, which add to the complexity of the H–T CH region. Comparison of the spectra in Figure 7.10 permits the identification of end-group resonances in PPO 4000, and by elimination, those resonances that are crosshatched must result from carbons in the H–H : T–T structure. These H–H : T–T peaks are identified as to carbon type from the DEPT spectra in Figure 7.9. A summary of the methine and methylene carbon resonances and their assignment to H–H : T–T defects or end groups appears in Table 7.4.

Table 7.4 ■ ^{13}C NMR Chemical-Shift Assignments and Relaxation Data
for the Methine and Methylene Carbons of Atactic PPO 4000
at 23°C[a]

Resonance	Chem. shift, ppm	Assignment[b]		T_1, sec
1	76.36	—CH—	E	0.93
2	76.29	—CH—	E	0.80
3	76.26	—CH$_2$—	3,4	0.80
4	76.21	—CH$_2$—	3,4	0.77
5	76.10	—CH$_2$—	E	1.19
6	75.98	—CH$_2$—	3,4	0.56
7	75.88	—CH$_2$—	3,4	0.59
8	75.24	—CH—	E	0.82
9	75.13	—CH—	E	0.81
10	75.08	—CH—, —CH$_2$—	E	0.81
11	75.02	—CH$_2$—	E	0.90
12	74.96	—CH—	E	1.08
13	74.91	—CH—	E	1.18
14	74.46	—CH$_2$—	2	1.04
15	74.26	—CH$_2$—	2	0.50
16	74.02	—CH$_2$—	2	0.51
17	73.82	—CH—	2,3	
18	72.97	—CH—	2,3	0.61
19	72.93	—CH—	2,3	0.64
20	72.87	—CH—	2,3	0.68
21	72.06	—CH$_2$—	E	2.16
22	72.03	—CH$_2$—	E	2.16
23	73.30	—CH—	2,3	
24	75.65	—CH$_2$—	E	
25	75.57	—CH$_2$—	E	

[a] Figure 7.9.
[b] E indicates chain end structure; 2,3,4 indicate H–H : T–T defect structure (see Table 7.2).

The methyl regions of both atactic PPO samples are displayed in Figure 7.11. Resonances attributable to methyl carbons in or adjacent to an end group can be assigned by comparison of (a) and (b), where it can be seen that all H–H : T–T defect resonances occur downfield from the H–T peaks. A summary of the methyl-carbon data is given in Table 7.5.

In order to make the assignments of the carbon nuclei in the H–H : T–T structures, a comparison was made between the experimental chemical-shift data (Figures 7.9 and 7.11 and Tables 7.4 and 7.5) and the relative chemical shifts for each carbon type resulting from the γ-*gauche*-effect calculations (Table 7.2). The calculated data indicate a lack of sensitivity for all carbons to the nature of the stereochemistry across the T–T portion of the chain (diad c

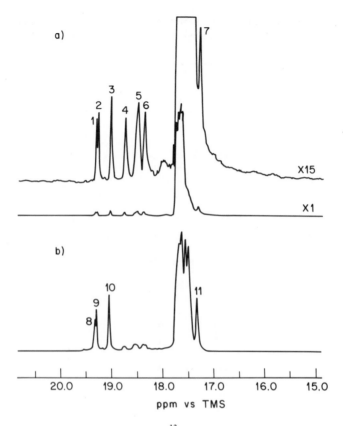

Figure 7.11 ▪ Methyl region of the 50.31-MHz ^{13}C NMR spectra of (a) atactic PPO 4000 and (b) atactic PPO 1000 observed at 23°C in C_6D_6. See Table 7.5. [Reprinted with permission from Schilling and Tonelli, (1986).]

in Table 7.2). In addition, for the H–H methyl and methylene carbons 2 and 3, diad *b* strongly affects their chemical shifts while diad *a* has a much smaller influence. The H–H methine carbons, however, are expected to show only a very minor stereochemical dependence.

Using the calculated shift data of Table 7.2, it is possible to make the specific assignments given in Tables 7.4 and 7.5. The H–H methine carbons 2 and 3 are predicted to be significantly upfield of the H–T CH resonances (at about 72.8–73.8 ppm) and are observed most clearly in the DEPT editing spectrum [Figure 7.9(b)]. Methine carbon 4 cannot be resolved from the H–T CH resonances.

For the CH$_2$ carbons [Figure 7.9(a)] the group of resonances slightly downfield of the H–T methylene resonances are assigned to carbon 2, while the methylene resonances (3, 4, 6, 7) shifted downfield into the H–T CH region

Table 7.5 ▪ ^{13}C NMR Chemical-Shift Assignments and Relaxation Data for the Methyl Carbons of Atactic PPO 4000 at 23°C^a

Resonance	Chem. shift, ppm	Assignment^b	T_1, sec
1	19.27	E	1.65
2	19.24	E	1.65
3	18.99	E	1.80
4	18.74	2, 3	0.99
5	18.51	2, 3	0.96
6	18.38	2, 3	0.92
7	17.29	E	1.09
8	19.31	E	
9	19.26	E	
10	19.02	E	
11	17.31	E	

^{a}Figure 7.11.
^{b}E indicates chain end structure; 2, 3 indicate H–H : T–T defect structure (see Table 7.2)

are assigned to carbons 3 and 4. Despite differences in the magnitudes of calculated and observed shifts for the H–H : T–T vs. H–T methine and methylene carbons (see below), the predicted direction for each carbon permits a consistent set of assignments as given in Table 7.4.

These results illustrate the difficulties faced by early workers in assigning the ^{13}C NMR spectrum of PPO. At first glance one is tempted to simply divide the 73–76-ppm region into two parts, methine and methylene. However, a careful interpretation of the chemical-shift effects produced by the H–H : T–T structures shows that a large number of methine and methylene resonances should overlap, and that the identity of carbon types can only be ascertained by DEPT or INEPT editing experiments (Turner, 1984). In addition, the comparison of PPO samples differing in molecular weight is necessary to identify the chain-end carbon resonances.

The differences in the magnitudes of the δ^{13}C's observed and calculated for methine and methylene carbons in the H–H : T–T and H–T structures may stem from the slightly different β-substituents (Stothers, 1972) present in each of these structural environments. The methine and methylene carbons in both H–T and H–H : T–T structures are β to oxygen ($CH-CH_2-O$ and CH_2-CH-O), and the methylene carbons are β to methyl carbons ($CH_2-CH-CH_3$). However, the H–T methine and H–H : T–T methylene carbons are β to methylene carbons ($CH-O-CH_2$ and CH_2-O-CH_2), while H–T methylene and H–H : T–T methine carbons are β to methine carbons (CH_2-O-CH and $CH-O-CH$) (see Table 7.2). On the other hand, H–H : T–T and H–T methyl carbons have precisely the same α and β substituents. The fact that the calculated and observed δ^{13}C's agree so closely for the methyl carbons (see below) lends support to the suggestion that slightly different β-substituents for H–T and H–H : T–T methine and methylene carbons may be the source of the disparity between the magnitudes of their measured and calculated δ ^{13}C's.

The methyl carbon resonances of the H–H : T–T structures are assigned in Table 7.5. Both the calculated magnitudes and the directions of the methyl resonances relative to the H–T resonances agree with the observed results [Figure 7.11(a)]. The three defect resonances (peaks 4, 5, 6) are assigned to carbons 2 and 3 (Table 7.2). Because of the predicted overlap resulting from stereosequences, we cannot make further specific assignments in this region.

Note that in Tables 7.3, 7.4, and 7.5 values of the spin–lattice relaxation time T_1 are presented for each resonance. These were determined by the inversion-recovery method (Farrar and Becker, 1971) and were utilized to insure that quantitative spectra were obtained. From the methyl-carbon data [Figure 7.11(a)] we are able to estimate that PPO 4000 contains 2.2% inverted or defect monomer units and has a number-average molecular weight M_n, based on the end-group resonance intensities observed, of 5400, or DP = 93. This may be compared with the $M_n = 4000$, or DP = 69, provided by the manufacturer and based on KOH hydroxyl number.

With the aid of multiple-pulse editing techniques (DEPT, INEPT) and the γ-gauche-effect calculations of relative ^{13}C NMR chemical shifts, we (Schilling and Tonelli, 1986) have assigned the ^{13}C NMR spectrum of PPO, including the determination of carbon resonances resulting from chain-end structures. Analysis of the expected differences between the γ-gauche interactions of the methine and methylene carbons in H–T and H–H : T–T PPO structures indicated that H–H : T–T methine resonances should overlap with the H–T methylene signals and that H–H : T–T methylene and H–T methine peaks should also overlap. It was this analysis that prompted our reinvestigation and assignment of the ^{13}C NMR spectrum of PPO. From these assignments it was possible to quantitatively determine the number of H–H : T–T defects incorporated in PPO through occasional ring opening of proylene oxide monomer at the CH—O bond. In addition, identification of chain-end structures permitted an estimate of the number-average molecular weight, and (though not discussed here) determination of specific terminal structures also provided insight concerning the polymerization mechanism of PPO.

References

Abe, A., Hirano, T., and Tsurata, T. (1979). *Macromolecules* **12**, 1092.

Bates, T. W. and Stockmayer, W. H. (1968). *Macromolecules* **1**, 12, 17.

Bax, A. (1982). *Two-Dimensional Nuclear Magnetic Resonance in Liquids*, Delft Univ. Press and Reidel Publ. Co., Dordrecht, Netherlands.

Bovey, F. A. (1982). *Chain Structure and Conformation of Macromolecules*, Academic Press, New York, Chapter 1.

Bovey, F. A. and Kwei, T. K. (1979). In *Macromolecules: An Introduction to Polymer Science*, F. A. Bovey and F. H. Winslow, Eds., Academic Press, New York, Chapter 3.

Bovey, F. A., Schilling, F. C., Kwei, T. K., and Frisch, H. L. (1977). *Macromolecules* **10**, 559.

Bruch, M. D., Bovey, F. A., and Cais, R. E. (1984). *Macromolecules* **17**, 1400.

Bruch, M. D., Bovey, F. A., Cais, R. E., and Noggle, J. H. (1985). *Macromolecules* **18**, 1253.

Cais, R. E. and Kometani, J. M. (1983). *Org. Coat. Appl. Polym. Sci. Proc. Am. Chem. Soc.* **48**, 216.

Cais, R. E. and Kometani, J. M. (1984). *Macromolecules* **17**, 1887.

Cais, R. E., and Kometani, J. M. (1985). *Macromolecules* **18**, 1354.

Cais, R. E. and Sloane, N. J. A. (1983). *Polymer (British)* **24**, 179.

Derome, A. E. (1987). *Modern NMR Techniques for Chemistry Research*, Pergamon, New York, Chapter 4.

Farrar, T. C. and Becker, E. D. (1971). *Pulse and Fourier Transform NMR*, Academic Press, New York.

Ferguson, R. C. and Brame, E. G., Jr. (1979). *J. Phys. Chem.* **83**, 1397.

Ferguson, R. C. and Ovenall, D. W. (1984). *Polym. Prepr. Am. Chem. Soc. Div. Polym. Chem.* **25**(1), 340.

Görlitz, M., Minke, R., Troutvetter, W., and Weisgerber, G. (1973). *Angew. Makromol. Chem.* **29/30**, 137.

Hirano, T., Khanh, P. H., and Tsurata, T. (1972). *Makromol. Chem.* **153**, 331.

Lapeyre, W., Cheradame, H., Spassky, N., and Sigwalt, P. (1973). *J. Chim. Phys.* **70**, 838.

Lyerla, J. R. and VanderHart, D. L. (1976). *J. Am. Chem. Soc.* **98**, 1697.

Oguni, N., Lee, K., and Tani, H. (1972). *Macromolecules* **5**, 819.

Oguni, N., Maeda, S., and Tani, H. (1973a). *Macromolecules* **6**, 459.

Oguni, N., Watanabe, S., Maki, M., and Tani, H. (1973b). *Macromolecules* **6**, 195.

Oguni, N., Shinohara, S., and Lee, K. (1979). *Polym. J. (Tokyo)* **11**, 755.

Price, C. C. and Osgan, M. (1956). *J. Am. Chem. Soc.* **78**, 4787.

Price, C. C., Spectro, R., and Tunolo, A. C. (1967). *J. Polym. Sci. Part A-1* **5**, 407.

Ramey, K. C. and Field, N. D. (1964). *Polym. Lett.* **2**, 461.

Schaefer, J. (1969). *Macromolecules* **2**, 533.

Schilling, F. C. (1982). *J. Magn. Reson.* **47**, 61.

Schilling, F. C. and Tonelli, A. E. (1980). *Macromolecules* **13**, 270.

Schilling, F. C. and Tonelli, A. E. (1986). **19**, 1337.

Stothers, J. B. (1972). *Carbon-13 NMR Spectroscopy*, Academic Press, New York, Chaps. 3 and 5.

Tani, H., Oguni, N., and Watanabe, S. (1968). *Polym. Lett.* **6**, 577.

Tonelli, A. E. (1976). *Macromolecules* **9**, 547.

Tonelli, A. E. and Schilling, F. C. (1981). *Macromolecules* **14**, 74.

Tonelli, A. E., Schilling, F. C., Starnes, W. H., Jr., Shepherd, L., and Plitz, I. M. (1979). *Macromolecules* **12**, 78.

Tonelli, A. E., Schilling, F. C., and Cais, R. E. (1981). *Macromolecules* **14**, 560.

Tonelli, A. E., Schilling, F. C., and Cais, R. E. (1982). *Macromolecules* **15**, 849.

Turner, C. J. (1984). *Prog. Nucl. Magn. Reson. Spectrosc.* **16**, 27.

Uryu, T., Shimazu, H., and Matsuzaki, K. (1973). *Polym. Lett.* **11**, 275.

Wilson, C. W., III. (1963). *J. Polym. Sci. Part A* **1**, 1305.

Wilson, C. W., III and Santee, E. R., Jr. (1965). *J. Polym. Sci. Part C* **8**, 97.

8

Copolymer Microstructure

8.1. Introduction

In our introductory discussion of polymer microstructures (Chapter 1) we mentioned that an important structural variable is provided by the ability to synthesize polymer chains composed of more than a single type of monomer unit. Copolymers, whose chains are composed of two or more different monomer units, may possess all the structural elements of homopolymers, i.e., stereo-, regio-, and geometrical isomerism, branching, etc., but in addition possess the structural variables of comonomer composition (amount) and sequence distribution (order). Possible copolymer types were illustrated in Chapter 1 (Section 1.3.5), where examples of random and regularly alternating copolymers were contrasted with block and graft copolymers.

Our discussion of copolymer microstructure will be limited to "random-type" copolymers where the comonomer units are intimately connected, or dispersed, along the copolymer chain. Though also scientifically and techno-logically interesting, block and graft copolymers, which contain relatively long sequences of one monomer bonded to similar long sequences of another monomer, will not be treated here, simply because their NMR spectra are virtually identical to their constituent homopolymer spectra. All we can reasonably expect to learn from their NMR spectra are the average lengths of the sequences containing a single monomer type.

Random copolymers, on the other hand, often exhibit NMR spectra that are significantly different from the spectra of their constituent homopolymers. This is a direct consequence of the structural intimacy between comonomer units resulting from a random comonomer sequence distribution. It is just this intimate comonomer sequence which makes NMR spectroscopy, a technique sensitive to the local molecular structures and conformations, so valuable to the determination of microstructures in random copolymers.

8.2. Comonomer Sequences

Let us consider how the comonomer units are distributed along a copolymer chain and how we can use NMR to determine the comonomer sequence. The copolymer made in the free-radical copolymerization of vinylidene chloride

$$
(V, \quad \underset{\underset{C\ell}{|}}{\overset{\overset{C\ell}{|}}{C}} = CH_2)
$$

and isobutylene

$$
(I, \quad \underset{\underset{CH_3}{|}}{\overset{\overset{CH_3}{|}}{C}} = CH_2)
$$

serves as a good example, because neither monomer generates asymmetric centers in the copolymer chain, which would lead to NMR spectra complicated by stereosequence effects. Furthermore, ^1H NMR may be effectively employed in the analysis of V–I copolymers, because of the absence of scalar ^1H–^1H J-coupling in this system.

Hellwege et al. (1966) and Kinsinger et al. (1966, 1967) first studied the vinylidene chloride–isobutylene (V–I) system by low-field (60 MHz) ^1H NMR. In Figure 8.1 the ^1H NMR spectrum of a V–I copolymer (65 mol% V) measured at 200 MHz by Bruch and Bovey (1984) is presented. Note the lines at the top of the spectrum corresponding to the resonances of the constituent poly(vinylidene chloride) and polyisobutylene homopolymers. The assignment of ^1H resonances to V–I comonomer tetrads (Bruch and Bovey, 1984) is presented in Table 8.1. This assignment is based on the earlier work of Hellwege et al. (1966) and Kinsinger et al. (1966, 1967) as verified and corrected by 2D NOE spectroscopy.

2D NOE spectroscopy (Bax, 1982) is a 2D NMR technique (see Sections 3.5 and 6.3) that produces a map of the entire network of nuclear Overhauser effects between the protons in a molecule. It is performed in much the same manner as 2D J-correlated spectroscopy (COSY) (see Sections 3.5, 6.3, and 7.2.3), except the correlating interaction is the nuclear Overhauser effect (see Sections 3.3 and 3.5) and not the scalar J-coupling between proton nuclei.

The 2D NOE pulse sequence is illustrated in Figure 8.2. Those protons not exchanging magnetization during the mixing period, which is about T_1, have the same initial and final frequency, giving rise to the normal spectrum (see Figure 7.1) along the diagonal. Those protons close enough (< 4 Å) to exchange magnetization due to dipole–dipole cross relaxation during τ_m have a final frequency different from the initial. This results in off-diagonal cross

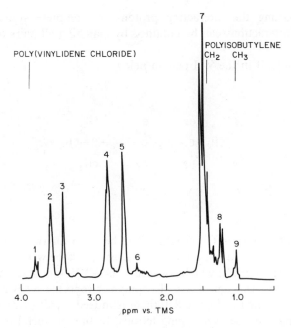

Figure 8.1 ▪ 200-MHz proton spectrum of a vinylidene chloride–isobutylene copolymer (65 mol% vinylidene chloride), observed at 40°C in a 20% solution in CDCl₃. The assignments are listed in Table 8.1 [Reprinted with permission from Bruch and Bovey (1984).]

Table 8.1 ▪ **Proton Resonance Assignments in Vinylidene Chloride–Isobutylene Copolymer Spectrum**[a]

Peak no.	Sequence assignment	Chem. shift vs. Me₄Si, ppm
1	VVVV CH₂	3.80
2	VVVI CH₂	3.58
3	IVVI CH₂	3.41
4	VVIV CH₂	2.80
5	VVII + IVIV CH₂	2.59
6	IVII CH₂	2.33
7	VIV CH₃	1.46
8	VII CH₃	1.24
9	III CH₃	1.03

[a](Bruch and Bovey (1984)).

peaks connecting the interacting protons. A complete system of nuclear Overhauser interactions can be obtained by matching all pairs of off-diagonal cross peaks.

For example, if in the V–I pentad below

$$
\begin{array}{ccccc}
C\ell & C\ell & C\ell & CH_3 & C\ell \\
| & | & | & | & | \\
-C-CH_2-C-CH_2-C-CH_2-C-CH_2-C- \\
| & | & | & | & | \\
C\ell & C\ell & C\ell & CH_3 & C\ell \\
\\
V & V & V & I & V
\end{array}
$$

there is a proton–proton Overhauser effect between the two central methylene groups, then we would expect to see a cross peak between the overlapping VVVI and VVIV tetrads, which constitute the VVVIV pentad. As can be seen in the 2D NOE spectrum of a V–I copolymer reported by Bruch and Bovey (1984) and presented in Figure 8.3, there is indeed a pair of cross peaks (2, 4) corresponding to these overlapping tetrads. In this manner Bruch and Bovey (1984) have made the complete comonomer sequence assignment of the V–I copolymer as presented in Figure 8.1 and in Table 8.1. It is apparent that this V–I copolymer has a "random" comonomer sequence in the sense that all ten possible tetrad comonomer sequences are observed. Later in this chapter (see Section 8.5) we will discuss an analysis of the V–I copolymerization mechanism made possible by the ^{1}H NMR determination of comonomer sequences just discussed.

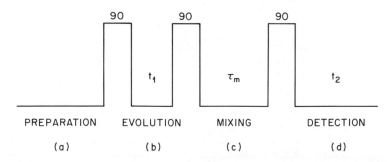

Figure 8.2 ■ 2D NOE pulse sequence: (a) system relaxes back to equilibrium; (b) protons are labeled by their initial precession frequencies; (c) dipolar-coupled spins exchange magnetization; (d) final precession frequencies are detected. [Reprinted with permission from Bruch and Bovey (1984).]

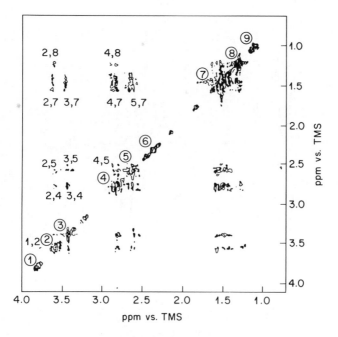

Figure 8.3 ■ Contour plot of 200-MHz 2D NOE spectrum of a vinylidene chloride–isobutylene copolymer (65 mol% vinylidene chloride) in $CDCl_3$. The temperature was 40°C and the mixing time 500 msec. The cross peaks indicate nuclear Overhauser effects between the indicated peaks. [Reprinted with permission from Bruch and Bovey (1984).]

8.3. Copolymer Stereosequences

Though propylene (P) does not homopolymerize under free-radical initiation (Deanin, 1967), it can be incorporated to a minor degree in a copolymerization with vinyl chloride (VC) leading to P–VC copolymers with up to 15 mol% P units. The combination of low P content and the inability of P to homopolymerize under these conditions results in P–VC copolymers where all P units are isolated by long uninterrupted runs of VC units. Consequently, we may study the stereochemistry of the comonomer sequences containing the isolated P units, i.e., $\cdots$ VC — VC — VC — VC — VC — P — VC — VC — VC — VC — VC $\cdots$, and compare them with the stereosequences found in the two homopolymers PP and PVC (see Chapters 5 and 6). This comparison may afford us the opportunity to gain some insight into the mechanism of the copolymerization of P and VC monomers.

A comparison of the ^{13}C NMR spectra of PVC and P–VC copolymer is made in Figure 8.4. From the ratio of P to VC methine-carbon resonance intensities measured in (b) we find (Tonelli and Schilling, 1984) that 5.1 mol% of the repeat units in this P–VC copolymer are P. The methyl carbon regions

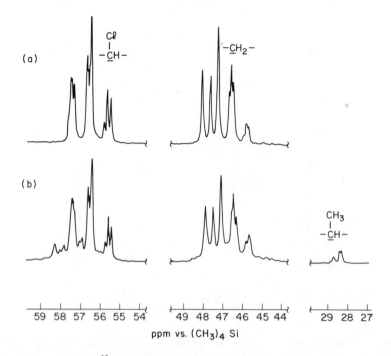

Figure 8.4 ■ (a) 50-MHz ^{13}C NMR spectrum of PVC. (b) VC and P methine and methylene carbon portions of the 50-MHz ^{13}C NMR spectrum of P–VC copolymer. [Reprinted with permission from Tonelli and Schilling (1984).]

of the ^{13}C NMR spectra of two PP samples are compared to the methyl carbon region of the P–VC copolymer in Figure 8.5. The PP sample (A) in (a) is a typical commercial atactic material (Schilling and Tonelli, 1980), while the PP sample (B) in (b) is a heptane-soluble fraction of a research-grade material (Plazek and Plazek, 1983). The stick spectrum in (b) was calculated as described in Chapter 6, while the γ-*gauche*-effect ^{13}C chemical shifts calculated for the methyl carbons in the P–VC copolymer (c) were obtained by employing Mark's (1973) RIS conformational model of P–VC copolymers.

Comparison of the methyl resonances in P–VC and PP reveals a decreased sensitivity to stereosequence for the P–VC copolymer. The methyl-carbon resonances in P–VC are sensitive to pentad stereosequences, while in PP heptad sensitivity is observed. In Table 8.2 the ^{13}C NMR chemical shifts calculated for the methyl carbons in several heptad stereosequences of P–VC and PP are compared. As observed, the methyl-carbon chemical shifts calculated for P–VC are sensitive to pentads, while PP methyl carbons show significant heptad sensitivity. This difference in stereosequence sensitivity between the methyl carbons in P–VC and PP is directly attributable to

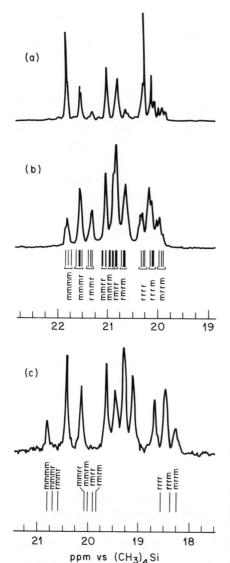

Figure 8.5 ■ (a) Methyl carbon region of the 50-MHz ^{13}C NMR spectrum of PP sample A. (b) Methyl carbon region of the 50-MHz ^{13}C NMR spectrum of PP sample B with stick spectrum of ^{13}C chemical shifts calculated for the methyl carbons in atactic PP. (c) P-methyl-carbon region of the 50-MHz ^{13}C NMR spectrum of P–VC copolymer with stick spectrum of ^{13}C chemical shifts calculated for the methyl carbons in P–VC. [Reprinted with permission from Tonelli and Schilling, (1984).]

differences in their conformational behavior as embodied in their RIS models. Local bond conformations reflect pentad sensitivity in P–VC and heptad dependence in PP. In addition, note that the overall spreads in methyl-carbon chemical shifts observed in P–VC and PP are 2.7 and 2.0 ppm, respectively, with the methyl carbons in P–VC resonating about 1 ppm upfield from those in PP. These observations are also reproduced by the calculated chemical

Table 8.2 ■ ^{13}C NMR Chemical Shifts Calculated for the Methyl Carbons in Several Heptad Stereosequences of P–VC and PP[a]

Heptad	$\Delta\delta$,[b] ppm	
	P–VC	PP
r(rmrm)r	0	0
m(rmrm)r	−0.01	−0.07
r(rmrm)m	−0.01	−0.05
m(rmrm)m	−0.03	−0.10
r(mrrm)r	0	0
m(mrrm)r	−0.04	−0.07
m(mrrm)m	−0.07	−0.12

[a] Tonelli and Schilling (1984).
[b] $\Delta\delta$ is the difference in chemical shift among the various heptads containing the same central pentad stereosequence. $\gamma_{CH_3,CH} = -5$ ppm was used for both PP and P–VC.

shifts, which employ the same γ-effect ($\gamma_{CH_3,CH} = -5$ ppm) and further indicate differences in the conformational behavior between P–VC copolymer (Mark, 1973) and PP homopolymer (Suter and Flory, 1975).

The methine region of the ^{13}C NMR spectrum of PP is shown in Figure 8.6(a). Part (b) of this figure presents the region of the P–VC ^{13}C NMR spectrum belonging to methine carbons in P units. Chemical shifts calculated for these methine carbons

$$\begin{array}{ccccc} C\ell & C\ell & CH_3 & C\ell & C\ell \\ | & | & |_* & | & | \\ \end{array}$$
$$(-C-C-C-C-C-\overset{*}{C}-C-C-C-C-C-)$$

are given in parts (c) and (d) of the same figure and correspond to $\gamma_{CH(CH_3),CH_2} = -5$ ppm and $\gamma_{CH(CH_3),Cl} = -7$ and -3 ppm, respectively. As we have seen in Chapter 5, ^{13}C NMR studies of PVC and its oligomers (Tonelli et al., 1979) indicate $\gamma_{CH(CH_3),Cl} = -3$ ppm, while comparisons of the ^{13}C NMR chemical shifts observed (Stothers, 1972) in alkanes and their chlorinated derivatives lead to $\gamma_{CH(CH_3),Cl} = -7$ ppm, as described in Chapter 4.

Comparison of the observed $CH(CH_3)$ resonances with the calculated chemical shifts shows a close correspondence for $\gamma_{CH(CH_3),Cl} = -7$ ppm [see Figure 8.6(b,c)], while $\gamma_{CH(CH_3),Cl} = -3$ ppm leads to calculated chemical shifts disparate from those observed [see Figure 8.6(b,d)]. This assignment is further supported by the observation that $P_m = P_r = 0.5$ from the methyl resonances (cf. seq.), leading to a 1 : 3 ratio of $mm : (mr + rr)$ centered pentad intensities, which is consistent with the chemical shifts calculated for the P methine carbons using $\gamma_{CH(CH_3),Cl} = -7$ ppm [see Figure 8.6(b,c)].

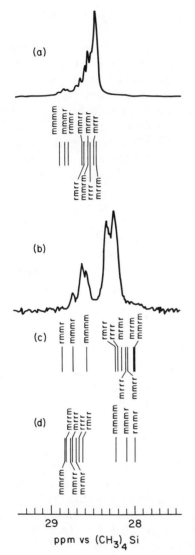

Figure 8.6 ■ (a) Methine-carbon region of the 50-MHz ^{13}C NMR spectrum of PP sample B with stick spectrum of ^{13}C chemical shifts calculated for PP shown below. (b) P-methine-carbon region of the 50-MHz ^{13}C NMR spectrum of P–VC copolymer. (c, d) Stick spectra of ^{13}C chemical shifts calculated for the P methine carbons in P–VC with $\gamma_{CH(CH_3),Cl} = -7$ and -3 ppm, respectively. [Reprinted with permission from Tonelli and Schilling (1984).]

If we focus on the regions of the P–VC ^{13}C NMR spectrum corresponding to the methyl and methylated methine carbons, it is possible to determine the statistics governing the stereochemical addition of a P unit to a long sequence of VC units and the addition of a VC unit to an isolated P unit. Integration of the ten CH_3 pentad peaks in Figure 8.5(c) permits an estimate of the concentration of $mm, mr(rm)$, and rr triads centered about the isolated P units. If the stereochemical placement is Bernoullian (see Chapter 6), then $P_m^2 = mm$, $P_r^2 = rr$, and $P_m + P_r = 1.0$. We find (Tonelli and Schilling, 1984)

$P_m = P_r = 0.5$, indicating that the addition of P to VC units and of VC to isolated P units is completely stereorandom.

This conclusion is supported by the intensities of the resonances in the methylated methine region of the ^{13}C NMR spectrum of P–VC in Figure 8.6(b), which also leads to $P_m = P_r = 0.5$ for the addition of P and the next VC unit. By contrast the long runs of VC units in PVC, though also described by Bernoullian statistics, favor the addition of racemic units with $P_r = 1 - P_m = 0.56$ [see Chapter 6 and Tonelli et al. (1979)]. From the area measurements of the methyl pentad resonances (Tonelli and Schilling, 1984) the stereochemistry of the first VC–VC diad following the P unit was found not to be Bernoullian, and therefore did not provide a P_m measure of the long VC runs in the P–VC copolymer.

Clearly, the stereochemical statistics for the copolymerization of P and VC are predominantly Bernoullian, as is observed for PVC homopolymerization (Tonelli et al., 1979). The homopolymerization of P, on the other hand, is not normally a Bernoullian process (Schilling and Tonelli, 1980; Inoue et al., 1984). Thus the addition of P and VC to VC is stereochemically Bernoullian, while the addition of P to P and of VC to P–VC to form the first VC–VC diad adjacent to the isolated P unit are not.

8.4. Copolymer Conformations

Copolymers formed in the free-radical copolymerization of styrene (S) and methyl methacrylate (MM) were among the first and most often studied by NMR (Bovey, 1962; Nishioka et al., 1962; Kato et al., 1964; Harwood, 1965; Harwood and Ritchey, 1965; Ito and Yamashita, 1965, 1968; Overberger and Yamamoto, 1965; Bauer et al., 1966; Ito et al., 1967; Yambumoto et al., 1970; Katritzky et al., 1974; Katritzky and Weiss, 1976; Yokota and Hirabayashi, 1976; Hirai et al., 1979; Koinuma et al., 1980; Heffner et al., 1986). As noted in Table 8.3, the number of possible structures in a copolymer, such as S–MM, where both monomers are capable of introducing an asymmetric center, is formidable. At the level of n consecutive comonomer units, the number $N'(n)$ of distinguishable structures is given by $N'(n) = 2^{2(n-1)} + 2^{n-1}$. Thus, there are 6, 20, 72, 272, and 1056 unique structures at the comonomer diad, triad, tetrad, pentad, and hexad levels.

^{13}C NMR spectra of S–MM copolymers show resolution of comonomer and stereosequences (Katritzky et al., 1974) superior to their ^{1}H NMR spectra, with the α-CH$_3$ resonance of the MM units and the aromatic C-1 resonance of the S unit

Table 8.3 ▪ Configurational Sequences in Copolymers[a]

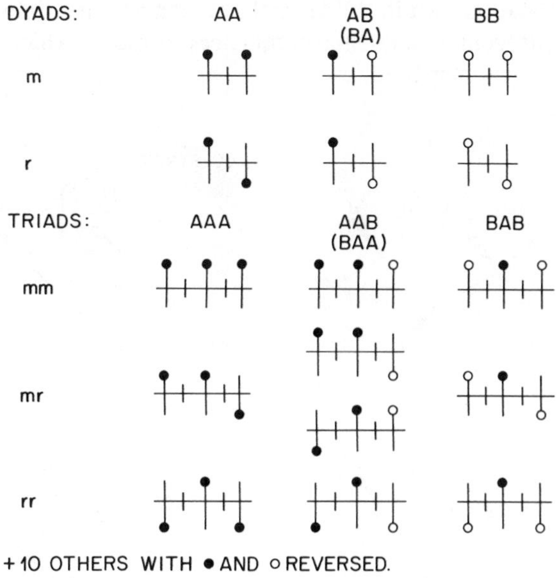

DYADS:	AA	AB (BA)	BB
m			
r			
TRIADS:	AAA	AAB (BAA)	BAB
mm			
mr			
rr			

+10 OTHERS WITH ● AND ○ REVERSED.

[a] Bovey (1982).

displaying the greatest sensitivity to tacticity. S–MM sequences were found to be nearly random in configuration, while runs or blocks of each monomer were predominantly syndiotactic. Both of these observations are consistent with the stereosequences observed in the NMR spectra of both homopolymers when produced by free-radical, or anionic, initiation. [More recently (Khanarian et al., 1982; Sato et al., 1982; Tonelli, 1983) it has been determined that polystyrene samples made by free-radical or anionic initiation are generally atactic.]

More interesting are the S–MM copolymers made from complexed monomers. Hirai et al. (1979) and Koinuma et al. (1980) observed that the S–MM copolymers prepared photochemically in the presence of the monomer-complexing agents stannic chloride or ethylaluminum sesquichloride showed NMR spectra quite different from the normal free-radical product. Figure 8.7 presents the 500-MHz ^{1}H NMR spectra of two 50:50 S–MM copolymers, one obtained by free-radical copolymerization and the other by copolymerization of the complexed monomer units. The broader resonances of the random 50:50 S–MM copolymer [see (a)] result from a large number of closely spaced, unresolved chemical shifts reflecting the large number of comonomer and stereosequences present in this copolymer. Restricting the comonomer sequence to the regularly alternating structure results in the spectral simplification seen in (b). Here the resonances are differentiated solely by the different comonomer stereosequences present, which are indicated below at the comonomer triad level:

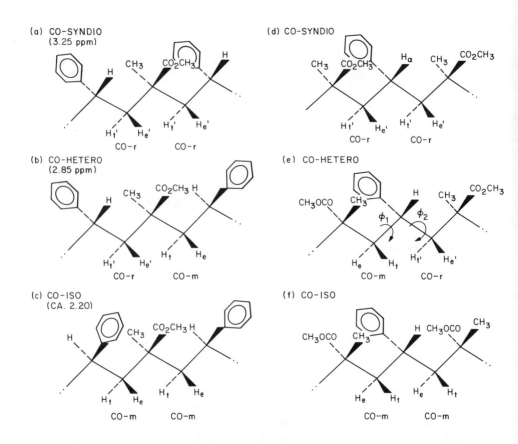

Let us illustrate how the 2D NMR technique NOESY, i.e. NOE correlated spectroscopy [see Section 8.2 and Wuthrich (1986)], can be applied to the regularly alternating S–MM copolymer to learn about its conformational characteristics. The nuclear Overhauser effect (NOE) depends on the direct through-space dipole–dipole interactions between nuclear spins (Noggle and Schirmer, 1971). During the mixing time, τ_m (see Figure 8.2), which is of the order of T_1 for the observed protons, spins that are close in space exchange magnetization by direct dipole–dipole interaction. Spins labeled by their frequencies during the evolution time, as in the usual J-correlated or COSY experiment (see Chapters 3 and 6), may precess at different frequencies at

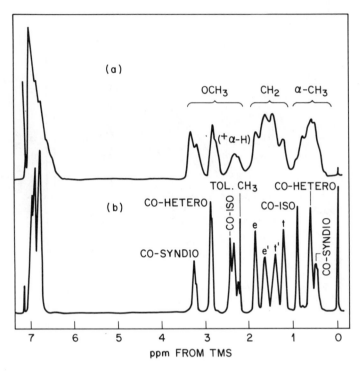

Figure 8.7 ▪ 500-MHz proton spectra of (a) the random and (b) alternating 50 : 50 styrene–methyl methacrylate copolymers, observed in 10% (w/v) hexachlorobutadiene solution at 80°C. [Reprinted with permission from Heffner et al. (1986).]

the end of τ_m. Cross peaks result between proton spins that are closer than about 4 Å.

The methylene region of the 500-MHz 2D-NOESY spectrum of the regularly alternating S–MM copolymer is presented in Figure 8.8 (Heffner et al., 1986). The stippled cross peaks correspond to the intermethylene interactions occurring in the S-centered co-hetero S–MM triad seen above in the triad drawing (e). These intermethylene NOE cross peaks appear to fall into three categories based on their intensities: one strong (S), $H_{e'}$–H_t; two medium (M), H_e–$H_{e'}$ and H_t–$H_{t'}$; and one weak (W), H_e–$H_{t'}$—corresponding to short, medium, and longer interproton distances, respectively. [See Heffner et al. (1986) for the details of proton-peak assignments.]

By combining portions of the conformational descriptions derived for styrene, methyl acrylate, and methyl methacrylate homopolymers (Yoon et al., 1975a, b; Sundararajan and Flory, 1974, 1977), Koinuma et al. (1980) developed a RIS model for the 1 : 1 alternating S–MM copolymer. When this RIS model is utilized to calculate the conformational probabilities for the bond pair flanking the styrene methine carbon in the co-hetero S–MM triad

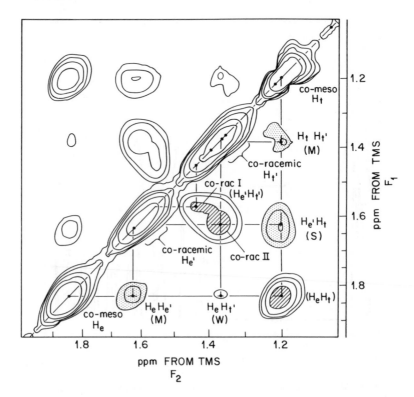

Figure 8.8 ▪ Expansion of the phase-sensitive proton NOESY spectrum at 500 MHz and 80°C, showing only the methylene region. Geminal interactions are indicated by hatched cross peaks, and intermethylene interactions by stippled cross peaks. The designations S, M, and W refer to the strengths of the intermethylene proton interactions. [Reprinted with permission from Heffner et al. (1986).]

illustrated above in drawing (e), it is possible to calculate (Heffner et al., 1986) the average intermethylene proton distances corresponding to the four cross peaks in Figure 8.8. This procedure is illustrated in Figure 8.9 and Table 8.4. The only three conformations allowed for the co-hetero S–MM triad are drawn in Figure 8.9 along with the probability calculated for each. The intermethylene proton distances, r_{HH}, calculated for the same S–MM triad are listed for each conformer in Table 8.4.

When these r_{HH}-values are raised to the -6 power and averaged over the three possible conformers shown in Figure 8.9 according to the calculated probabilities also listed there, the entries in the next-to-last row of Table 8.4 are obtained. These values should be proportional to the strengths of the intermethylene proton–proton cross peaks seen in Figure 8.8, and this is indeed the case.

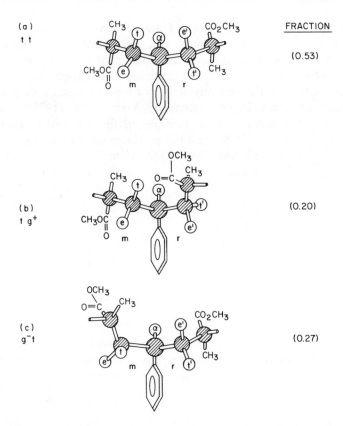

Figure 8.9 ■ Ball-and-stick models of the styrene-centered MM–S–MM co-hetero triad, showing the tt, tg[+], and g[−]t conformations, with 20° deviations from exact staggering. [Reprinted with permission from Heffner et al. (1986).]

Table 8.4 ■ **Intermethylene H–H Distances (r_{HH}) Calculated for the Co-hetero Styrene-Centered Triad [see Figure 8.9 and Structure (e)]**

	r_{HH}, Å			
ϕ_1, ϕ_2	$H_e–H_{t'}$	$H_e–H_{e'}$	$H_t–H_{t'}$	$H_{e'}–H_t$
t, t $(-20°, 20°)$	2.89	3.10	3.10	2.20
t, g[+] $(-20°, 100°)$	3.74	2.63	3.68	2.59
g[−], t $(-100°, 20°)$	3.74	3.68	2.63	2.59
$\langle \phi_1, \phi_2 \rangle$	0.0010[a]	0.0016[a]	0.0016[a]	0.0063[a]
$\langle \phi_1, \phi_2 \rangle$	(0.0027)[b]	(0.0016)[b]	(0.0020)[b]	(0.0030)[b]

[a] r_{HH}^{-6} averaged over all three (ϕ_1, ϕ_2) conformations.
[b] Same as above, except $\phi_1, \phi_2 = 0, \pm120°$ in the t, g[±] states (Heffner et al., 1986).

The agreement between the predicted and observed pattern of NOESY cross peaks for the co-hetero triad of $1:1$ alternating S–MM copolymer confirms the validity of the Koinuma et al. (1980) conformational model. It is particularly noteworthy that this agreement requires the assumption of about $20°$ displacements from the perfectly staggered rotational states as predicted for the backbone bonds in polystyrene by Yoon et al. (1975a) (see the Newman projections below). As an example, in the t, t conformation (see Figure 8.9) $\phi_1, \phi_2 = -20°, 20°$ because this produces relief from steric interactions of the phenyl ring and the methyl methacrylate C_α, as can be seen in the following Newman projections:

If perfectly staggered states t $(0°)$, $g^\pm$ $(\pm 120°)$ are assigned in the calculation of intermethylene proton–proton distances, then the results in the bottom row (in parentheses) of Table 8.4 are obtained, i.e., all interactions $(\langle r_{HH}^{-6} \rangle)$ are approximately the same. It is apparent from Figure 8.8 that this is not the case.

More recently Mirau et al. (1987) have derived intermethylene proton–proton distances r_{HH} directly from the NOESY spectra of $1:1$ alternating S–MM. A comparison was made with the distances r_{HH} of Table 8.4 after conformational averaging according to the Koinuma et al. (1980) RIS model. Reasonable agreement was obtained, as indicated by the following observed and (calculated) conformer populations: $t, t = 0.58 \pm 0.05$ (0.53), $t, g^+ = 0.24 \pm 0.05$ (0.20), and $g^-, t = 0.18 \pm 0.05$ (0.27). In addition, it was found that an $11°$ displacement from perfectly staggered rotational states produced r_{HH}-values in closest agreement with those obtained from NOESY cross-peak intensities, adding further support to the Koinuma et al. (1980) RIS model, which assumed $20°$ displacements. This 2D NOESY ^{1}H NMR study of $1:1$ alternating S–MM copolymer marked the first attempt to derive the conformational characteristics of a flexible polymer in solution through a direct measure of conformationally averaged interproton distances.

8.5. Copolymerization Mechanisms

During copolymerization of two comonomers it is often observed that the monomers are not incorporated into the resultant copolymer chain in the same proportion as the initial comonomer mixture. This is a result of the differing and competing reactivities of the monomers and the growing chain ends. Mayo and Lewis (1944) have shown that the relationship between the instantaneous copolymer composition and monomer feed composition (starting ratio of comonomers) is described by

$$\frac{d[M_1]}{d[M_2]} = \frac{[M_1]}{[M_2]} \frac{r_1[M_1] + [M_2]}{r_2[M_2] + [M_1]}, \tag{8.1}$$

where M_1 and M_2 are the two comonomers. The ratio of the rates at which the two comonomers enter the copolymer, or the instantaneous copolymer composition, is given by the left-hand side of this equation. The mole ratio of monomers in the feed is just $[M_1]/[M_2]$. The quantities r_1 and r_2 are the comonomer reactivity ratios. If k_{11} is the rate constant for the addition of M_1 to a growing chain ending in an M_1 unit, and k_{12} is the rate constant describing the addition of M_2 to a growing chain also terminated by an M_1 unit, then

$$r_1 = k_{11}/k_{12}, \tag{8.2}$$

$$r_2 = k_{22}/k_{21}, \tag{8.3}$$

where k_{22} and k_{21} are the rate constants describing the addition of M_2 and M_1 to a growing chain terminated with a M_2 unit.

From the copolymer equation 8.1, the instantaneous copolymer composition is obtained as

$$F_1 = 1 - F_2 = \frac{d[M_1]}{d[M_1] + d[M_2]}. \tag{8.4}$$

In terms of the mole feed ratios f_1 and f_2 for monomers M_1 and M_2, i.e. $f_1 = 1 - f_2 = [M_1]/([M_1] + [M_2])$,

$$F_1 = \frac{r_1 f_1^2 + f_1 f_2}{r_1 f_1^2 + 2 f_1 f_2 + r_2 f_2^2}, \tag{8.5}$$

where f_1, f_2 and F_1, F_2 are the mole fractions of M_1, M_2 in the feed or reaction mixture and in the copolymer, respectively.

It is important to stress again that the copolymerization equation describes the instantaneous copolymer composition. In most copolymerizations the comonomers are not incorporated into the copolymer in the same proportion as the feed composition, resulting in a changing feed composition depleted in the more reactive monomer. Consequently, as the monomer conversion increases, the copolymer composition becomes increasingly heterogeneous. This makes it more difficult to interpret the microstructure of the copolymer in terms of the relative reactivity ratios. For this reason most studies of copolymerization limit the monomer conversion to less than about 5%.

Reactivity ratios provide important information concerning the interactions of comonomers with their growing copolymer chains, which constitutes the mechanism of their copolymerization. Traditionally their determination has required the evaluation of the overall composition of copolymers prepared from a series of different feed ratios. Elemental analysis is then generally used to obtain the overall comonomer compositions.

A variety of computational and graphical methods have been derived to extract values of the reactivity ratios from the overall copolymer composition data. As an example, if the copolymer equation in the form of Eq. 8.5 is

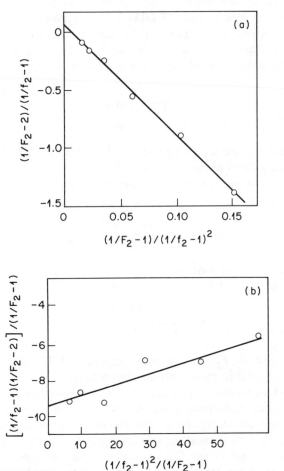

Figure 8.10 ■ Fineman and Ross's (1950) method for determining copolymer reactivity ratios for the system N-vinylsuccinimide(1)–methyl acrylate(2) [Hopff and Schlumbom (1977)]: $r_1 = 0.07$ (a) and $r_2 = 9.3$ (b). [Reprinted with permission from Bovey (1982).]

Table 8.5 ▪ Reactivity Ratios for Copolymerization at 60°C[a]

M_1	M_2	r_1	r_2
Styrene	Acrylonitrile	0.4 ± 0.05	0.04 ± 0.04
Styrene	Methyl methacrylate	0.52 ± 0.02	0.46 ± 0.02
Styrene	Butadiene	0.78 ± 0.01	1.39 ± 0.03
Styrene	Vinyl acetate	55 ± 10	0.01 ± 0.01
Styrene	Maleic anhydride	0.02	0
Methyl methacrylate	Acrylonitrile	1.2 ± 0.14	0.15 ± 0.07
Methyl methacrylate	Vinyl acetate	20 ± 3	0.015 ± 0.015
Methyl methacrylate	Methyl acrylate	1.69	0.34
Vinyl acetate	Acrylonitrile	0.061 ± 0.013	4.05 ± 0.3
Vinyl acetate	Vinyl chloride	0.23 ± 0.02	1.68 ± 0.08
Vinylidene chloride	Isobutene	3.3	0.05

[a] Bovey (1982).

rearranged to

$$\frac{1/F_2 - 2}{1/f_2 - 1} = r_1 - r_2 \frac{1/F_2 - 1}{(1/f_2 - 1)^2} \tag{8.6}$$

and then plotted as shown in Figure 8.10(a), it is possible (Fineman and Ross, 1950) to obtain r_1 from the intercept on the ordinate. The data may also be plotted according to

$$\frac{(1/f_2 - 1)(1/F_2 - 2)}{1/F_2 - 1} = -r_2 + r_1 \frac{(1/f_2 - 1)^2}{1/F_2 - 1}, \tag{8.7}$$

from which r_2 is obtained [see Figure 8.10(b)]. This analysis indicates that methyl acrylate (2) is much more reactive toward its own growing chain radicals $(r_2 = k_{22}/k_{21} = 9.3)$ than is N-vinylsuccinimide(1) $(r_1 = k_{11}/k_{12} = 0.07)$. Table 8.5 summarizes reactivity ratios obtained for the free-radical copolymerization of several comonomer pairs.

The graphical and computational derivation of reactivity ratios from overall copolymer composition data is at best tedious and often crude and insensitive. A better approach, made possible by the advent of NMR analysis, is to measure the frequencies of comonomer sequences, because the same theoretical treatment of copolymerization that predicts overall comonomer compositions also predicts the frequencies of occurrence of specific comonomer sequences. Deviations from the "terminal" model, which assumes that only the end monomer unit of the growing copolymer chain determines its reactivity, can be detected by the determination of comonomer sequences. In addition, reactivity ratios can be obtained from the comonomer sequence frequencies determined for a single copolymer, provided the feed ratio used in its copolymerization is known.

In the random copolymerization of two monomers free of asymmetric centers, diad, triad, and tetrad comonomer sequences may be represented as follows:

Diads: m_1m_1 m_1m_2 (or m_2m_1) m_2m_2
Triads: $m_1m_1m_1$ $m_2m_2m_2$
          $m_1m_1m_2$ (or $m_2m_1m_1$) $m_1m_2m_2$ (or $m_2m_2m_1$)
          $m_2m_1m_2$ $m_1m_2m_1$
Tetrads: $m_1m_1m_1m_1$ $m_1m_1m_2m_1$ ($m_1m_2m_1m_1$) $m_2m_2m_2m_2$
          $m_1m_1m_1m_2$ ($m_2m_1m_1m_1$) $m_1m_1m_2m_2$ ($m_2m_2m_1m_1$) $m_2m_2m_2m_1$ ($m_1m_2m_2m_2$)
                                      $m_2m_1m_2m_1$ ($m_1m_2m_1m_2$)
          $m_2m_1m_1m_2$ $m_2m_1m_2m_2$ ($m_2m_2m_1m_2$) $m_1m_2m_2m_1$

Their frequencies of occurrence, or diad probabilities, are given by

$$[m_1m_1] = F_1P_{11},\tag{8.8}$$

$$[m_1m_2]\ (\text{or } [m_2m_1]) = 2F_1P_{12} = 2F_1(1 - P_{11}),\tag{8.9}$$

$$= 2F_2P_{21} = 2F_2(1 - P_{22}),\tag{8.10}$$

$$[m_2m_2] = F_2P_{22},\tag{8.11}$$

where F_1 and F_2 are again the overall mole fractions of comonomers m_1 and m_2, and P_{11}, P_{12}, P_{21}, and P_{22} are the probabilities corresponding to the rate constants k_{11}, k_{12}, k_{21}, and k_{22}. For instance, the probability that monomer m_2 will add to the growing copolymer chain terminated by an m_1 end unit is given by P_{12}. The probabilities P_{11} and P_{22} can also be shown to be expressible in terms of monomer-feed mole fractions (f_1, f_2) and reactivity ratios by

$$P_{11} = \frac{r_1f_1}{1 - f_1(1 - r_1)},\tag{8.12}$$

$$P_{22} = \frac{r_2f_2}{1 - f_2(1 - r_2)},\tag{8.13}$$

from which

$$r_1 = \frac{(1 - f_1)[m_1m_1]}{f_1(F_1 - [m_1m_1])},\tag{8.14}$$

$$r_2 = \frac{(1 - f_2)[m_2m_2]}{f_2(F_2 - [m_2m_2])}.\tag{8.15}$$

Analogous relations apply to triad and tetrad comonomer sequences.

In Section 8.2 we presented a discussion of the 1D and 2D ^{1}H NMR analysis of comonomer sequences in the vinylidene chloride (V or m_1): isobutylene (I or m_2) copolymer. The relative intensity of the $[m_1m_1]$ (VV)-centered resonances between 3.4 and 3.8 ppm (see Figure 8.1 and Table 8.1) is

0.426. According to Eq. 8.14, r_1 must be 3.31. Because $[m_2m_2]$ (II) is not directly obtainable from the NMR spectrum due to overlap, we may instead use the group of $[m_1m_2]$ (VI) resonances near 2.8 ppm to evaluate r_2. Clearly

$$[m_1m_2](+[m_2m_1]) = 2F_2(1 - P_{22}), \qquad (8.16)$$

or

$$[m_1m_2](+[m_2m_1]) = 2F_2 - \frac{2F_2r_2f_2}{1 - r_2} \qquad (8.17)$$

The relative intensity of $[m_1m_2]$ (VI) resonances leads via Eq. 8.17 to $r_2 = 0.04$.

Kinsinger et al. (1966) obtained $r_1 = k_{11}/k_{12} = 3.3$ and $r_2 = k_{22}/k_{21} = 0.05$ by conventional analysis of overall V–I composition data. Within experimental error, both approaches lead to the same reactivity ratios. Vinylidene chloride radicals prefer to add vinylidene chloride, while chains terminated by isobutylene units rarely add another isobutylene.

Consideration of comonomer tetrad frequencies (Kinsinger et al., 1967) reveals that the reactivity of the growing copolymer chain terminated with vinylidene chloride depends on whether the penultimate unit is another vinylidene chloride unit or an isobutylene unit. It was found that if the penultimate unit is isobutylene, the copolymer chain is twice as likely to add a vinylidene chloride unit to form ----IVV · .

An additional important advantage of the comonomer sequence determination of copolymerization mechanisms by NMR is the capacity to treat the copolymerization of asymmetric monomers, such as the styrene–methyl methacrylate system discussed in Section 8.4. Traditional methods are powerless when it comes to treating asymmetric copolymers, because they are based on overall comonomer compositions, which do not permit differentiation between comonomer stereosequences. As an example, the interesting stereosequence information concerning the P–VC copolymer with isolated P units, as revealed by ^{13}C NMR analysis in Section 8.3, is not obtainable through the traditional approaches. In this connection it should be mentioned that the reactivity ratios presented in Table 8.5 do not convey any information concerning the comonomer stereosequences occurring in these asymmetric copolymers.

References

Bauer, R. G., Harwood, H. J., and Ritchey, W. M. (1966). *Polym. Preprints* **7** (2), 973.

Bax, A. (1982). *Two-Dimensional Nuclear Magnetic Resonance in Liquids*, Delft University Press (Delft) and D. Reidel (Amsterdam).

Blouin, F. A., Cheng, R. C., Quinn, M. H., and Harwood, H. J. (1973). *Polym. Preprints* **14** (1), 25.

Bovey, F. A. (1962). *J. Polym. Sci.* **62**, 197.

Bovey, F. A. (1982). *Chain Structure and Conformation of Macromolecules*, Academic Press, New York, Chapter 5.

Bruch, M. D. and Bovey, F. A. (1984). *Macromolecules* **17**, 978.

Deanin, R. D. (1967). *SPE J.* **23** (5), 50.

Fineman, M. and Ross, S. D. (1950). *J. Polym. Sci.* **5**, 259.

Harwood, H. J. (1965). *Angew. Chem. Int. Ed. Eng.* **4**, 1051.

Harwood, H. J. and Ritchey, W. M. (1965). *J. Polym. Sci. Part B* **3**, 419.

Heffner, S. A., Bovey, F. A., Verge, L. A., Mirau, P. A., and Tonelli, A. E. (1986). *Macromolecules* **19**, 1628.

Hellwege, K. H., Johnsen, U., and Kolbe, K. (1966). *Kolloid-Z.* **214**, 45.

Hirai, H., Tanabe, T., and Koinuma, H. (1979). *J. Polym. Sci. Polym. Phys. Ed.* **17**, 843.

Hopff, H. and Schlumbom, P. C. (1977). Cited in Elias, H. G. (1977). *Macromolecules*, Plenum, New York, p. 768.

Inoue, Y., Itabashi, Y., Chujo, R., and Doi, Y. (1984). *Polymer (British)* **25**, 1640.

Ito, K. and Yamashita, Y. (1965). *J. Polym. Sci. Part B* **3**, 625, 631.

Ito, K. and Yamashita, Y. (1968). *J. Polym. Sci. Part B* **6**, 227.

Ito, K., Iwase, S., Umehara, K., and Yamashita, Y. (1967). *J. Macromol. Sci. Part A* **1**, 891.

Kato, Y., Ashikari, N., and Nishioka, A. (1964). *Bull. Chem. Soc. Jpn.* **37**, 1630.

Katritzky, A. R. and Weiss, D. E. (1976). *Chem. Brit.* **45**.

Katritzky, A. R., Smith, A., and Weiss, D. E. (1974). *J. Chem. Soc. Perkin Trans.* **2**, 1547.

Khanarian, G., Cais, R. E., Kometani, J. M., and Tonelli, A. E. (1982). *Macromolecules* **15**, 866.

Kinsinger, J. B., Fischer, T., and Wilson, C. W., III (1966). *J. Polym. Sci. Part B* **4**, 379.

Kinsinger, J. B., Fischer, T., and Wilson, C. W., III (1967). *J. Polym. Sci. Part B* **5**, 285.

Koinuma, H., Tanabe, T., and Hirai, H. (1980). *Makromol. Chem.* **181**, 383.

Mark, J. E. (1973). *J. Polym. Sci. Polym. Phys. Ed.* **11**, 1375.

Mayo, F. R. and Lewis, F. M. (1944). *J. Am. Chem. Soc.* **66**, 1594.

Mirau, P. A., Bovey, F. A., Tonelli, A. E., and Heffner, S. A. (1987). *Macromolecules* **20**, 1701.

Nishioka, A., Kato, Y., and Ashikari, N. (1962). *J. Polym. Sci.* **62S**, 10.

Noggle, J. H. and Shirmer, R. E. (1971). *The Nuclear Overhauser Effect*, Academic Press, New York.

Overberger, C. G. and Yamamoto, N. (1965). *J. Polym. Sci. Part B* **3**, 569.

Plazek, D. L. and Plazek, D. J. (1983). *Macromolecules* **16**, 1469.

Sato, H., Tanaka, Y., and Hatada, K. (1982). *Makromol. Chem., Rapid Commun.* **3**, 175, 181.

Schilling, F. C. and Tonelli, A. E. (1980). *Macromolecules* **13**, 270.

Stothers, J. B. (1972). *Carbon-13 NMR Spectroscopy*, Academic Press, New York.

Sundararajan, P. R. and Flory, P. J. (1974). *J. Am. Chem. Soc.* **96**, 5025.

Sundararajan, P. R. and Flory, P. J. (1977). *J. Polym. Sci. Polym. Lett. Ed.* **15**, 699.

Suter, U. W. and Flory, P. J. (1975). *Macromolecules* **8**, 765.

Tonelli, A. E. (1983). *Macromolecules* **16**, 609.

Tonelli, A. E. and Schilling, F. C. (1984). *Macromolecules* **17**, 1946.

Tonelli, A. E., Schilling, F. C., Starnes, W. H., Jr., Shepherd, L., and Plitz, I. M. (1979). *Macromolecules* **12**, 78.

Wuthrich, K. (1986). *NMR of Proteins and Nucleic Acids*, Wiley, New York.

Yabumoto, S., Ishi, K., Arita, K., and Arita, K. (1970). *J. Polym. Sci. Part A-1* **8**, 295.

Yokota, K. and Hirabayashi, T. (1976). *J. Polym. Sci. Polym. Chem. Ed.* **17**, 57.

Yoon, D. Y., Sundararajan, P. R., and Flory, P. J. (1975a). *Macromolecules* **8**, 776.

Yoon, D. Y., Suter, U. W., Sundararajan, P. R., and Flory, P. J. (1975b). *Macromolecules* **8**, 784.

Chemically Modified Polymers

9.1. Introduction

The modification of polymer microstructures by means of postpolymerization chemical reactions is an increasingly active branch of polymer science. Specialty polymers may thus be obtained through the chemical modification of readily available polymers. Unique polymer microstructures, many of which cannot be achieved directly through homo- or copolymerization, can be created via this route. We may introduce functional or reactive groups into polymers, alter their surface structures, provide grafts and unusual side-chain substituents, and assist their analytical characterization by chemical modification. The radiation chemistry of polymers, which can result in their crosslinking to form networks and/or in their degradation via chain scission, is also a form of chemical modification.

In this chapter we describe the chemical modification of two common polymers, poly(vinyl chloride) (PVC) and poly(butadiene) (PBD). NMR spectroscopy is employed to determine the microstructures resulting from the chemical modifications, which lead to conclusions concerning the mechanisms of these polymer modifying chemical reactions. In one case, the dechlorination of PVC with the reducing agent tri-n-butyltin hydride, NMR studies of the polymer and its oligomeric model compounds provide kinetic information concerning the rates of their dechlorination.

9.2. Transformation of PVC to Ethylene–Vinyl Chloride Copolymers

9.2.1. Tri-n-butyltin Hydride Reduction of PVC

Traditional means of obtaining ethylene–vinyl chloride (E–V) copolymers suffer from several shortcomings. Direct copolymerization of E and V monomers does not usually lead to random E–V copolymers covering the entire range of comonomer composition. Free-radical copolymerization at low pressure (Misono et al., 1967, 1968) yields E–V copolymers with V contents from 60 to 100 mol%. The γ-ray-induced copolymerization under high pressure (Hagiwara et al., 1969) yields E–V copolymers with increased amounts of E, but it appears difficult to achieve degrees of E incorporation greater than 60 mol% without producing blocky samples. Chlorination of polyethylene (Keller and Mugge, 1976) results in head-to-head (vicinal) and multiple (geminal) chlorination, which are structures not characteristic of E–V copolymers.

Reduction of poly(vinyl chloride) (PVC) with tri-n-butyltin hydride [(n-Bu)$_3$SnH] (Starnes et al., 1983),

$$-CH_2-\underset{\underset{Cl}{|}}{CH}-CH_2-\underset{\underset{Cl}{|}}{CH}-CH_2- + (n-Bu)_3\,SnH \longrightarrow -CH_2-CH_2-CH_2-\underset{\underset{Cl}{|}}{CH}-CH_2- + (n-Bu)_3\,SnCl$$

on the other hand, is found to produce "random" E–V copolymers covering the complete composition range from PVC to polyethylene (PE) (Schilling et al., 1985). The reductive dechlorination of PVC with (n-Bu)$_3$-SnH produces E–V copolymers with the same chain length as the starting PVC, and their microstructures are readily determined by ^{13}C NMR analysis (Schilling et al., 1985) as indicated in Figure 9.1.

9.2.2. Microstructures of E–V Copolymers

The analysis of E–V microstructures rests on our ability to assign the resonances observed in their ^{13}C NMR spectra to specific comonomer and stereosequences. This is achieved by the calculation of ^{13}C chemical shifts for the various E–V microstructures according to the γ-*gauche*-effect method (Tonelli and Schilling, 1981). Using the RIS model developed by Mark (1973) for E–V copolymers and the γ-*gauche* effects determined in our studies of PVC and its oligomeric model compounds [Tonelli et al. (1979) and see Chapter 5] —i.e., $\gamma_{CH \text{ or } CH_2,CH_2} = -5$ ppm, $\gamma_{CH_2,CH} = -2.5$ ppm, and $\gamma_{CH \text{ or } CH_2,Cl} = -3$ ppm— ^{13}C chemical shifts were calculated for the E–V microstructures presented in Figure 9.2. Note that methylene carbons in certain E–V microstructures, (e.g. compare E, LVC1, and VCD methylenes) may possess 0, 1, or 2 chlorine atoms in the β-position. From ^{13}C NMR studies of chlorinated paraffins (Stothers, 1972) it is observed that a single chlorine substituent in the β-position produces a downfield shift of $+10.5$ ppm, while two β-Cl's result in a deshielding of $+19.5$ ppm. In the evaluation of ^{13}C chemical shifts for E–V copolymers both the γ-*gauche* and β-Cl effects were considered (see Chapter 4). Figure 9.3 presents a comparison of the ^{13}C chemical shifts observed and predicted for several E–V copolymers.

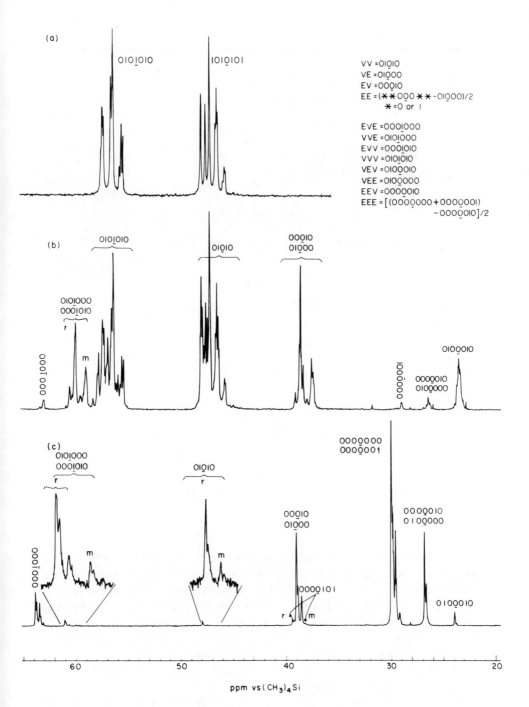

Figure 9.1 ■ 50.31-MHz ^{13}C NMR spectra of PVC (a) and two partially reduced PVCs, E–V-84 (b) and E–V-21 (c). Note the table of E–V microstructural designations in the upper right-hand corner, where $0,1 \equiv CH_2, CHCl$ carbons. Resonances correspond to underlined carbons. The assignment of different stereosequences is given by Schilling et al. (1985). [Reprinted with permission from Schilling et al. (1985).]

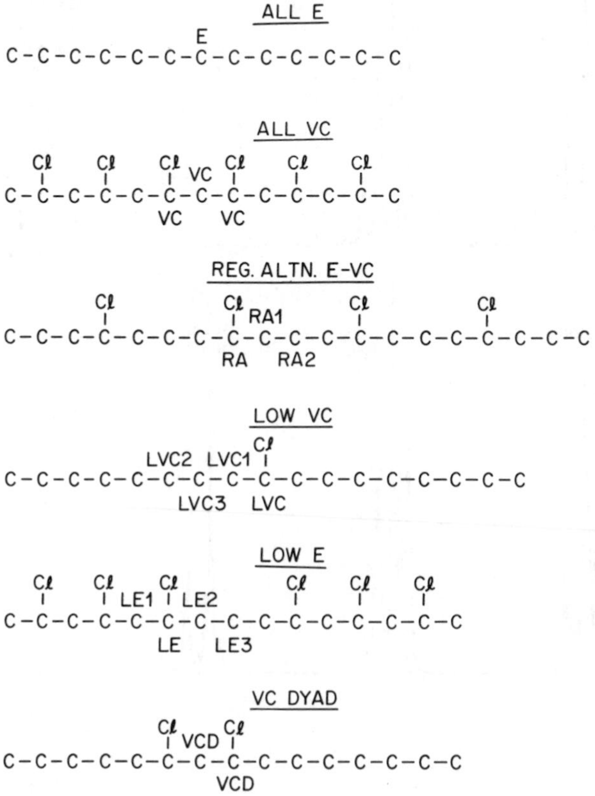

Figure 9.2 ■ Possible microstructures in E–V copolymers. [Reprinted with permission from Tonelli and Schilling (1981).]

The agreement between observed and calculated ^{13}C chemical shifts in E–V copolymers confirms the γ-effects derived from the ^{13}C NMR spectra of PVC and its oligomers (Tonelli et al., 1979) and the conformational model of E–V copolymers developed by Mark (1973). Our ability to predict the ^{13}C chemical shifts for the various microstructures occurring in E–V copolymers enables us to make the assignment of resonances indicated in Figure 9.1. Integration of the areas of these resonances permits the evaluation of comonomer sequence probabilities, as well as the overall comonomer composition of E–V copolymers. These results are summarized in Table 9.1.

In Table 9.2 we present a comparison of the observed comonomer diad and triad fractions with those expected for the completely random removal of chlorine from the starting PVC by the $(n\text{-Bu})_3$SnH reducing agent. For lightly reduced E–V copolymers, such as E–V-84, the observed distribution of comonomer sequences is nearly random. However, as the reduction proceeds it

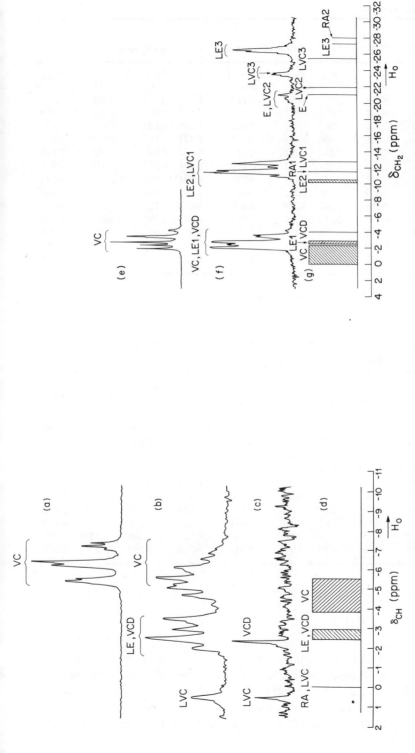

Figure 9.3 ■ (a) 25.16-MHz ^{13}C NMR spectrum (methine-carbon region) of atactic PVC observed at 107°C, using a 10% (w/v) solution in 1,2,4-trichlorobenzene. Addition of 62.5 ppm yields chemical shifts relative to Me$_4$Si. (b) Same as (a) except 29% thiol-reduced PVC (Starnes et al., 1978). (c) Same as (a) except 99.6% LiAlH$_4$-reduced PVC (Tonelli and Schilling 1981). (d) ^{13}C chemical shifts calculated at 100°C for the methine carbons residing in the E–VC microstructures of Figure 9.2. The widths of the resonances result from chemical-shift dispersion produced by the stereoregularity of VC units. (e) Same as (a) except methylene-carbon region. (f) Same as (b) except methylene-carbon region. (g) ^{13}C chemical shifts calculated at 100°C for the methylene carbons residing in the E–VC microstructures of Figure 9.2. The widths of the resonances result from chemical-shift dispersion produced by the stereoregularity of VC units. [Reprinted with permission from Tonelli and Schilling (1981).]

143

Table 9.1 ▪ Diad and Triad Probabilities of E–V Copolymers[a]

Copolymer[b]	P_{VV}	$P_{VE} = P_{VE}$	P_{EE}	P_{EVE}	$P_{VVE} = P_{EVV}$	P_{VVV}	P_{VEV}	$P_{VEE} = P_{EEV}$	P_{EEE}
E–V-85	0.742	0.124	0.011	0.015	0.115	0.619	0.114	0.011	0.0
E–V-84	0.709	0.134	0.023	0.025	0.108	0.615	0.101	0.019	0.004
E–V-71	0.470	0.239	0.052	0.063	0.175	0.310	0.175	0.048	0.008
E–V-62	0.344	0.278	0.099	0.116	0.177	0.177	0.177	0.075	0.027
E–V-61	0.343	0.275	0.107	0.121	0.173	0.198	0.141	0.083	0.029
E–V-60	0.316	0.285	0.114	0.141	0.167	0.154	0.179	0.077	0.038
E–V-50	0.200	0.297	0.205	0.192	0.133	0.073	0.166	0.129	0.045
E–V-46	0.147	0.309	0.235	0.205	0.116	0.037	0.149	0.140	0.098
E–V-37	0.087	0.286	0.342	0.219	0.078	0.012	0.115	0.158	0.183
E–V-35	0.061	0.278	0.383	0.224	0.064	0.015	0.090	0.168	0.208
E–V-21	0.014	0.197	0.593	0.190	0.016	0.0	0.035	0.153	0.436
E–V-14	0.0	0.127	0.746	0.104	0.0	0.0	0.051	0.123	0.599
E–V-2	0.0	0.025	0.950	0.021	0.0	0.0	0.0	0.026	0.926

[a]Schilling et al., (1985).
[b]The designation E–V-62, for example, signifies 62 mol% V units.

Table 9.2 ▪ Comparison of Observed Diad and Triad Fractions with Those Calculated for a Random or Bernoullian Distribution of Comonomers[a]

Copolymer	X_V	Dyad fractions			Triad fractions					
		VV	VE + EV	EE	VVV	VVE + EVV	VEV	VEE + EEV	EVE	EEE
E–V-84	84.3	0.709[b]	0.268	0.023	0.615	0.216	0.101	0.038	0.025	0.004
		0.711[c]	0.266	0.023	0.599	0.223	0.111	0.042	0.021	0.004
E–V-62	62.3	0.344	0.556	0.100	0.177	0.354	0.177	0.150	0.116	0.027
		0.388	0.470	0.142	0.242	0.292	0.146	0.177	0.089	0.054
E–V-46	45.6	0.147	0.618	0.235	0.037	0.231	0.149	0.280	0.205	0.098
		0.208	0.496	0.296	0.095	0.226	0.113	0.270	0.135	0.161
E–V-21	21.2	0.014	0.394	0.592	0.000	0.032	0.035	0.306	0.191	0.436
		0.045	0.334	0.621	0.010	0.071	0.035	0.263	0.132	0.489
E–V-14	13.6	0.000	0.254	0.746	0.000	0.000	0.051	0.246	0.104	0.599
		0.018	0.236	0.746	0.003	0.032	0.016	0.203	0.101	0.645

[a]Schilling et al. (1985).
[b]Observed.
[c]Random or Bernoullian.

becomes apparent from the observed diad and triad fractions that chlorines belonging to V units adjacent to other V units are preferentially removed relative to those adjacent to E units. This results in a decrease in the number of all-V and all-E unit runs from those expected from the random removal of chlorine atoms and leads to an enhanced alternation of E and V units as reflected by the greater than expected amounts of EVE and VEV triads.

Somewhere between 80 and 85% reduction all VV diads are removed, and all V units are flanked by at least one E unit on both sides.

In part (c) of Figure 9.1 the ^{13}C NMR spectrum of E–V-21 is presented together with the microstructural assignment of resonances (Schilling et al., 1985) based on comparison with the ^{13}C chemical shifts calculated via the γ-*gauche*-effect method [see Figure 9.3 and Tonelli and Schilling (1981)] and to the spectra of chlorinated *n*-alkane model compounds (Schilling et al. 1985). The methine carbon in the central V unit of the VVE and EVV (0101000 and 0001010) triads shows two groups of resonances centered at 60.4 ppm which correspond to racemic (*r*) and *meso* (*m*) VV diads. On the basis of the observation that the methine carbons in *meso*-2,4-dichloropentane and *meso*-4,6-dichlorononane resonate ~ 1 ppm upfield from their racemic isomers (Schilling et al., 1985), in agreement with their calculated γ-*gauche*-effect chemical shifts (Tonelli and Schilling, 1981), we assign the upfield resonance to *m*-(VVE + EVV) triads and the more intense downfield resonance to *r*-(VVE + EVV) triads.

The resonances between 47.4 and 48.0 ppm correspond to the methylene carbons surrounded by methine carbons in VV (01010) diads with the upfield peak assigned to the *m*-VV diad and the downfield peak to the *r*-VV diad, again in agreement with the order of resonances observed in the dichloroalkanes (Schilling et al., 1985). As expected, the integral area of the VV methylene carbon resonances is half the area of the methine resonances in the VVE + EVV triads, and the ratio of racemic to *meso* intensities is the same (4.2) in both regions.

The unreduced PVC is a Bernoullian polymer with $P_m = 0.44$ and, therefore, a ratio of *r* to *m* diads of $P_r/P_m = 0.56/0.44 = 1.27$ [see Figure 9.1(a)]. Removal of 79% of the chlorine atoms by $(n\text{-Bu})_3$SnH reduction results in E–V-21 copolymer with an *r* to *m* diad ratio of 4.2. Clearly *m*-diads are preferentially reduced by $(n\text{-Bu})_3$SnH, as was also observed in the LiAlH$_4$ reductions of PVC (Starnes et al., 1979). In the LiAlH$_4$ reductions of PVC, isolated VV diads were observed even after 98% of the chlorines were removed. This is in contrast to the PVC reduction with $(n\text{-Bu})_3$SnH, where VV diads disappear between 80 and 85% reduction. Consequently, isolated VV diads are more readily reduced by $(n\text{-Bu})_3$SnH than by LiAlH$_4$.

In Table 9.3 we present the observed ratios of *r* to *m* diads as a function of the degree of reduction for the starting PVC and five E–V copolymers [four obtained by $(n\text{-Bu})_3$SnH reduction (Schilling et al., 1985) and one by reduction with LiAlH$_4$ (Starnes et al., 1979)]. As noted by Starnes et al. (1979), we can obtain the ratio of rate constants (k_m/k_r) governing the reduction of *m* and *r* VV diads by comparing the concentrations of *m* and *r* diads observed in the starting PVC with those remaining in the E–V copolymer. For the four $(n\text{-Bu})_3$SnH reduced E–V copolymers $k_m/k_r = 1.32 \pm 0.1$, while for the LiAlH$_4$ reduced E–V, Starnes et al. (1979) found a similar value $k_m/k_r = 1.48$.

Table 9.3 ▪ VV-Dyad Stereosequence Composition of E–V Copolymers[a]

E–V copolymer	Cl removed, mol%	r/m[b]	k_m/k_r
PVC	0	1.27	
E–V-46	54.4	1.9	1.21
E–V-37	62.7	2.8	1.38
E–V-35	65.2	2.8	1.32
E–V-21	78.8	4.2	1.33
E–V-2[c]	98.1	11.1	1.48

[a]Schilling et al. (1985).
[b]Racemic/*meso* ratio.
[c]Obtained through $LiAlH_4$ reduction by Starnes et al. (1979).

It would appear that the ratio of rate constants governing the reduction of *m* and *r* diads is independent of the degree of reduction. Apparently the relative rates of chlorine removal from *m* and *r* VV diads is independent of longer-range E–V copolymer microstructure, such as which particular parent E–V triad contains the VV diad.

9.2.3. (*n*-Bu)₃SnH Reduction of PVC Model Compounds

It is possible to conclude from the ^{13}C NMR analysis of E–V copolymers that in the $(n\text{-}Bu)_3SnH$ reduction of PVC to E–V copolymers, and eventually to PE, chlorines belonging to VV diads are preferentially removed relative to isolated chlorines (EVE), and that *m*-VV diads are reduced faster than *r*-VV diads. Jameison et al. (1986, 1988) studied the $(n\text{-}Bu)_3SnH$ reduction of the PVC diad and triad model compounds 2,4-dichloropentane (DCP) and 2,4,6-trichloroheptane (TCH) in the hope that they would serve as useful models for the reduction of PVC to E–V copolymers. Unlike the polymers (PVC and E–V), DCP and TCH are low-molecular-weight liquids whose high-resolution ^{13}C NMR spectra can be recorded in a matter of minutes. Thus, it was possible to monitor their $(n\text{-}Bu)_3SnH$ reduction directly in the NMR tube and follow the kinetics of their dechlorination.

At 50°C Jameison et al. (1986, 1988) found that the $(n\text{-}Bu)_3SnH$ reduction of the PVC model compounds DCP and TCH reached 80% completion after 5 hours. The percentage reduction was determined by comparing the amounts of $(n\text{-}Bu)_3SnH$ and $(n\text{-}Bu)_3SnCl$ at each measurement point. It was necessary to record 10 scans at each sampling point as the reduction proceeded, with a 30-sec delay between each scan. Spin–lattice relaxation times T_1 were measured for each carbon in DCP, TCH, $(n\text{-}Bu)_3SnH$, $(n\text{-}Bu)_3SnCl$, 2-chloropentane, pentane, 2,4-dichloroheptane, 2,6-dichloroheptane, 2-chloroheptane,

4-chloroheptane, and heptane to verify that the 30-sec delay time between scans was sufficient to obtain quantitative spectra (Freeman and Hill, 1971).

As illustrated below, DCP(D) is sequentially transformed into 2-chloropentane (M) and then to pentane (P) during its reduction by $(n\text{-Bu})_3\text{SnH}$:

$$D \xrightarrow{k_D} M \xrightarrow{k_M} P.$$

The ratio of rate constants $K = k_M/k_D$ can be obtained (Benson, 1960) from the concentrations of D and M measured at various degrees of reduction, x, according to

$$\frac{M_x}{D_x} = \frac{1 - (D_x/D_0)^{K-1}}{K - 1}, \tag{9.1}$$

where the subscripts 0 and x indicate concentrations initially and after $x\%$ reduction.

An alternative means to determine the relative rates of reduction of M and D, i.e. $K = k_M/k_D$, is afforded by comparing the simultaneous $(n\text{-Bu})_3\text{SnH}$ reduction of DCP and 2-chlorooctane (M') to pentane (P) and octane (O), respectively.

$$D \xrightarrow{k_D} M \xrightarrow{k_M} P,$$

$$M' \xrightarrow{k_{M'}} O.$$

In this case $K' = k_{M'}/k_D$ is given by (Benson, 1960)

$$K' = \frac{\ln(M'_x/M'_0)}{\ln(D_x/D_0)}. \tag{9.2}$$

Equation 9.2 can also be used to determine the relative rates of reduction of *meso* (m) and racemic (r) DCP (D_m, D_r), where M' and D are replaced by D_m and D_r:

$$D_m \xrightarrow{k_{D_m}} M \xrightarrow{k_M} P,$$

$$D_r \xrightarrow{k_{D_r}} M \xrightarrow{k_M} P.$$

In the early stages of the reduction of TCH (T) with $(n\text{-Bu})_3\text{SnH}$ it is possible to compare the relative reactivities of the central (4) and terminal (2,6) chlorines. At these early levels of reduction only 2,6- and 2,4-dichloroheptanes (2,6-D and 2,4-D) are produced, as shown below:

$$T \xrightarrow{k_C} 2,6\text{-D},$$

$$T \xrightarrow{k_T} 2,4\text{-D}.$$

We can establish the relative reactivities, k_C/k_T, of the central (C) and terminal (T) chlorines directly from the relative concentrations of the resulting dichloroheptanes:

$$\frac{k_C}{k_T} = \frac{2,6\text{-D}}{2,4\text{-D}}. \tag{9.3}$$

The portion of the 50.13-MHz ^{13}C NMR spectra containing the methylene and methine carbon resonances of DCP and the resultant products of its $(n\text{-Bu})_3$SnH reduction are presented in Figure 9.4 at several degrees of reduction. Comparison of the intensities of resonances possessing similar relaxation times T_1 (see above) permits a quantitative accounting of the amounts of each species (D, M, P) present at any degree of reduction.

In Figure 9.5 the percentages of D (DCP), M (2-chloropentane), and P (pentane) observed during the $(n\text{-Bu})_3$SnH reduction of DCP are plotted against the degree of reduction x. Equation 9.1 is solved for $K = k_M/k_D$ by least-squares fitting the calculated and observed values of the ratio M_x/D_x. The observed ratios D_x/D_0 are substituted into Eq. 9.1 to obtain the calculated ratios M_x/D_x corresponding to the assumed $K = k_M/k_D$, and these are compared with the observed ratios M_x/D_x. This procedure yields $K = k_M/k_D = 0.26$, which means that DCP is about 4 times more easily reduced than 2-chloropentane (M). Comparison of the simultaneous reduction of DCP and 2-chlorooctane (M') gave, according to Eq. 9.2, $K' = k_{M'}/k_D = 0.24$, lending further support to the observation that chlorines belonging to a VV diad are removed 4 times faster than an isolated chlorine in (say) an EVE triad. Furthermore, the observed rates of $(n\text{-Bu})_3$SnH reduction of 2- and 4-chlorooctanes were identical within experimental error, indicating that the reactivity of an isolated chlorine is independent of structural position or chain-end effects.

The observed ratios of m to r isomers, m/r, remaining during the $(n\text{-Bu})_3$SnH reduction of DCP (see Figure 9.4) are plotted in Figure 9.6. Substituting these data into Eq. 9.2 yields a ratio $k_m/k_r = 1.3$. Apparently m-VV diads are 30% more reactive than are r-VV diads.

^{13}C NMR analysis of E–V copolymers obtained via the $(n\text{-Bu})_3$SnH reduction of PVC (Schilling et al., 1985) led to $k_m/k_r = 1.31 + 0.1$, in excellent

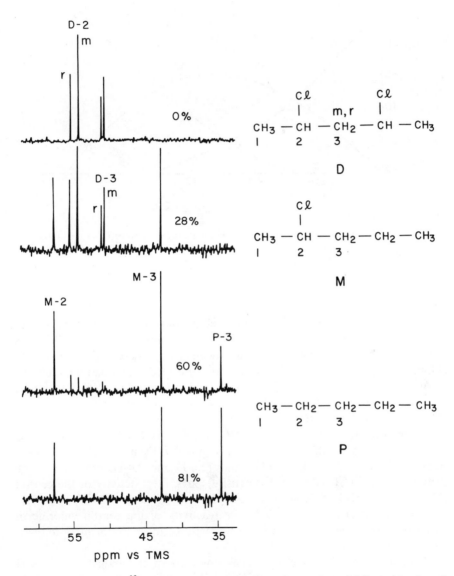

Figure 9.4 ■ 50.13-MHz ¹³C NMR spectra of DCP (D) and its products (M, P) resulting from 0, 28, 60, and 81% reduction with (*n*-Bu)₃SnH. [Reprinted with permission from Jameison et al. (1986, 1988).]

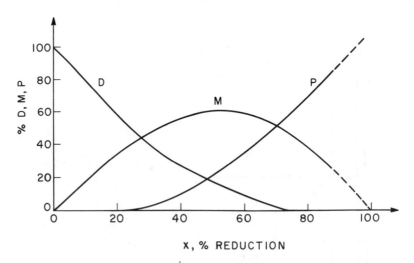

Figure 9.5 ■ Distribution of reactants (D, M) and products (M, P) observed in the $(n\text{-Bu})_3\text{SnH}$ reduction of DCP. [Reprinted with permission from Jameison et al. (1986, 1988).] D = DCP, M = 2-chloropentane, and P = pentane.

agreement with the kinetics observed for the removal of chlorines from m- and r-DCP (Jameison et al., 1986, 1988). It was also observed that no VV diads exist in those E–V copolymers made by removing more than 80% of the chlorines from PVC. This observation is confirmed and quantified by the $(n\text{-Bu})_3\text{SnH}$ reduction of DCP, where the chlorines in this PVC diad model compound were found to be 4 times easier to remove than the isolated chlorines in 2-chloropentane, and 2- and 4-chlorooctane.

The methine regions of the ^{13}C NMR spectra of TCH before and after 43% reduction with $(n\text{-Bu})_3\text{SnH}$ are shown in Figure 9.7. From the relative concentrations of 2,6- and 2,4-dichloroheptanes (2,6-D and 2,4-D) observed in the early stages of TCH reduction with $(n\text{-Bu})_3\text{SnH}$, the reactivity of the central chlorine in TCH is 50% greater than that of the terminal chlorines, i.e., $k_C/k_T = 1.5$ according to Eq. 9.3. The reactivity of the central chlorine in TCH was also found to depend on its stereoisomeric environment as follows: $mm > mr$ or $rm > rr$.

Comparison of the triad sequences observed in the reduction of TCH (Jameison et al., 1986, 1988) and PVC (Schilling et al., 1985) with $(n\text{-Bu})_3\text{SnH}$ is made in Figure 9.8 as a function of the percentage reduction. The curves numbered 0, 1, 2, and 3 correspond to triads containing 0 (EEE), 1 (EEV + VEE + EVE), 2 (VVE + EVV + VEV), and 3 (VVV) chlorine atoms. Agreement between the curves describing the products of reduction for TCH and PVC provides strong evidence for considering TCH an appropriate model

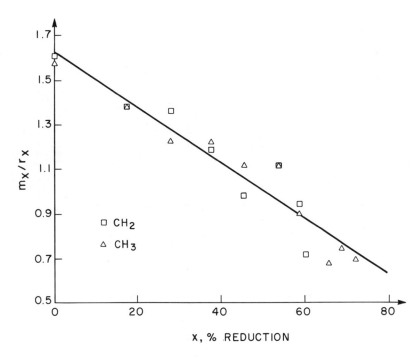

Figure 9.6 ▪ Ratio of the amounts of m and r isomers of DCP remaining after reduction by $(n\text{-Bu})_3\text{SnH}$, as measured by the ^{13}C methylene (see Figure 9.4) and methyl resonances. [Reprinted with permission from Jameison et al. (1986, 1988).]

compound for the $(n\text{-Bu})_3\text{SnH}$ reduction of PVC, and clearly implies that the $(n\text{-Bu})_3\text{SnH}$ reduction of PVC is independent of comonomer sequences longer than triads.

The first column of Table 9.4 lists all possible ^{13}C-NMR-distinguishable E–V triads whose central units are V. The same triad structures in binary notation ($0 = \text{E}$, $1 = \text{V}$) are presented in the second column, with the central V unit labeled as the site of $(n\text{-Bu})_3\text{SnH}$ attack and the terminal units as either $-$ (preceding site) or $+$ (following site). Relative reactivities of the central V (1) unit in each triad toward $(n\text{-Bu})_3\text{SnH}$ based on the kinetics of reduction determined for DCP and TCH are presented in the final column. For EVV (011) triads, removal of the central chlorine is expected to be 3.5 (r) and 4.6 (m) times faster than for the isolated chlorine atom in the EVE (010) triad, because $k_D/k_M = 4.0$ and $k_m/k_r = 1.3$ for DCP. The central chlorine in VVV (111) triads are 6.0 (mr or rm), 4.6 (rr), and 7.8 (mm) times more reactive toward $(n\text{-Bu})_3\text{SnH}$ than the chlorine in EVE, based on $k_D/k_M = 4.0$ and $k_m/k_r = 1.3$ for DCP and $k_C/k_T = 1.5$ for TCH.

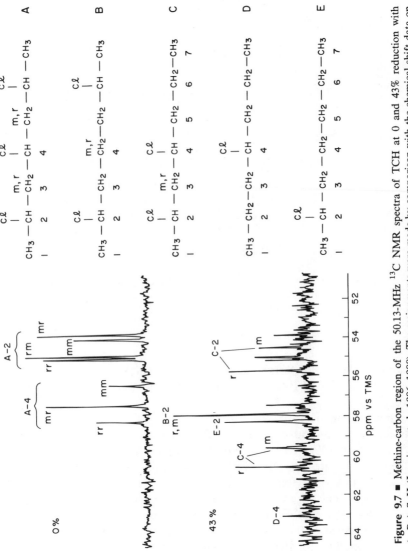

Figure 9.7 ■ Methine-carbon region of the 50.13-MHz ^{13}C NMR spectra of TCH at 0 and 43% reduction with $(n\text{-Bu})_3$SnH (Jameison et al., 1986, 1988). The assignments were made by comparison with the chemical-shift data on TCH (Tonelli et al., 1979), DCP, and 2- and 4-chlorooctane. [Reprinted with permission from Jameison et al. (1986, 1988).]

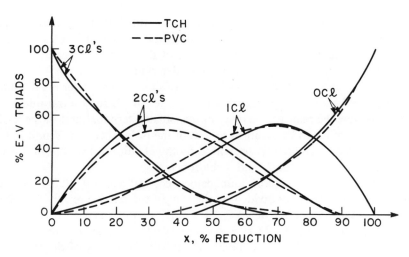

Figure 9.8 ▪ Comonomer triad distributions observed by ^{13}C NMR analysis during the $(n\text{-}Bu)_3SnH$ reductions of TCH (solid curves) (Jameison et al., 1986, 1988) and PVC (dashed curves) (Schilling et al., 1985). [Reprinted with permission from Jameison et al. (1986, 1988).]

Table 9.4 ▪ Relative Reactivities of the Central Chlorines in E–V Triads[a]

E–V Triad	Reduction site				k (relative)
	$-$			$+$	
EVE	0		1	0	1.0
EVV	0		1 r	1	3.5
EVV	0		1 m	1	4.6
VVV	1 m		1 r	1	$4.0 \times 1.5 = 6.0$
VVV	1 r		1 m	1 r	$6.0/1.3 = 4.6$
VVV	1		1 m	1	$6.0 \times 1.3 = 7.8$

[a]Jameison et al. (1986, 1988).

9.2.4. Computer Simulation of TCH and PVC Reduction

We begin with 100 TCH molecules reflecting the stereochemical composition of the unreduced TCH sample, i.e. 52 (mr or rm), 28 (rr), and 20 (mm) (Jameison et al., 1986, 1988). Selection of TCH molecules is based on the generation of random integers I_r, where $I_r < 101$. If $I_r < 53$, then the TCH molecule chosen is an mr or rm isomer. If $52 < I_r < 81$, then the TCH is rr, and if $I_r > 80$ it is mm.

Next we randomly choose either one or the other terminal units or the central unit of the selected TCH isomer and check to see if it is a V or E unit. If a terminal V unit is chosen, we check to see if the neighboring central unit is V or E. If the central unit is also V, then we determine whether this VV diad is *m* or *r*. For *r* and *m* VV diads the relative reactivities of the terminal-V-unit chlorines are 3.5 and 4.6, respectively (see Table 9.4). If the central unit is E, then we assume the relative reactivity of the isolated terminal unit in the VEE, EEV, or VEV triads to be identical to that of the isolated central V unit in the EVE triad. This assumption is supported by the identical rates of $(n\text{-Bu})_3\text{SnH}$ reduction observed for the 2- and 4-chlorooctanes.

Finally, we select a random number between 0.0 and 1.0. If it is smaller than the relative reactivity divided by the sum of the relative reactivities of all chlorines in the VVV, EVV or VVE, VEV, VEE or EEV, and EVE isomers of TCH and partially reduced TCH (see Table 9.4), then we remove the terminal chlorine (V → E, or 1 → 0) and modify the relative reactivity of the central V unit in the selected TCH isomer, because its terminal neighbor has been changed from V to E (1 to 0).

This procedure is repeated until the desired percentage reduction x is reached, where $x = 100 \times (\text{number of chlorines removed})/300$. Each TCH molecule is then tested for the number and sequence of V units remaining at the current value of x. In Figure 9.9 the simulated percentages of TCH molecules containing 3, 2, 1 and 0 chlorines, or V units, are plotted and compared with the values observed for various degrees of $(n\text{-Bu})_3\text{SnH}$ reduction. Agreement between the simulated and observed reduction products of

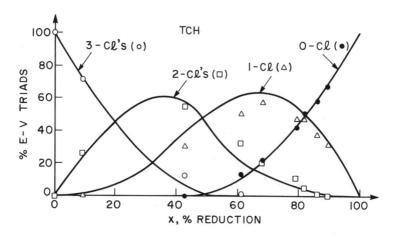

Figure 9.9 ■ Comparison of the observed (symbols) and simulated (solid lines) comonomer triad distributions in $(n\text{-Bu})_3\text{SnH}$-reduced TCH. [Reprinted with permission from Jameison et al. (1986, 1988).]

TCH based on the kinetics of reduction observed for both DCP and TCH is good.

Simulation of the $(n\text{-Bu})_3$SnH reduction of PVC is carried out in a manner similar to that just described for TCH. Instead of beginning with 100 TCH molecules, we take a 1000-repeat-unit PVC chain that has been Monte Carlo generated (Tonelli and Valenciano, 1986) to reproduce the stereosequence composition of the experimental sample of PVC used in the reduction to obtain E–V copolymers (Schilling et al., 1985), i.e. a Bernoullian PVC with $P_m = 0.44$. At this point the generated PVC chain is identical in length (molecular weight) and stereochemical structure to the experimental starting sample of PVC.

Repeat units are selected at random, and if they are unreacted V units, a check of whether or not the units adjacent to the selected unit are E or V is made. Having determined the triad structure (both comonomer and stereosequence) of the repeat unit selected for reduction, the relative reactivity of this E–V triad is divided by the sum of relative reactivities for all V-centered E–V triads, as listed in Table 9.4, to obtain the probability of reduction. A random number between 0.0 and 1.0 is generated, and if it is smaller than the probability of reduction of the selected E–V triad, the chlorine is removed from the central V unit, which becomes an E unit.

If either of the terminal units of the E–V triad selected for reduction is a V unit, then its relative reactivity is modified to reflect changing the central unit from V to E. The degree of reduction is calculated from $100 \times$ (number of chlorines removed)/1000, and if it corresponds to the desired level of reduc-

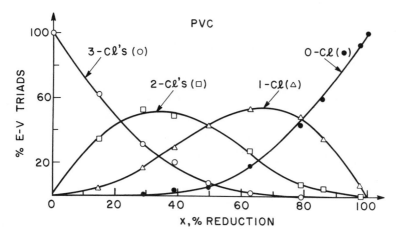

Figure 9.10 ■ Comparison of the observed (symbols) and simulated (solid lines) comonomer triad distributions in $(n\text{-Bu})_3$SnH-reduced PVC. [Reprinted with permission from Jameison et al. (1986, 1988).]

tion, the numbers of each type of triad remaining in the E–V copolymer are printed. The entire procedure is repeated for several PVC chains until the fraction of each triad type at each degree of reduction remains constant when averaged over the generated set of PVC chains.

A comparison of the observed (see Table 9.1) and simulated E–V triad compositions is presented in Figure 9.10 against the degree of overall reduction by $(n\text{-Bu})_3\text{SnH}$. The agreement is excellent, being much improved over that found for TCH reduction. This is at least partially a consequence of the relative accuracy of the ^{13}C NMR data used to obtain the E–V triad compositions resulting from the reduction of PVC, because the TCH data (Jameison et al., 1986, 1988) are gathered during the reduction and are an average over the time required to accumulate ^{13}C NMR spectra (about 10 minutes), while the E–V data (Schilling et al., 1985) are obtained on static samples removed from the reaction flask.

Ratios r/m for VV diads observed by ^{13}C NMR in E–V copolymers obtained by the $(n\text{-Bu})_3\text{SnH}$ reduction of PVC are compared in Figure 9.11 with those resulting from the computer simulation of PVC reduction based on the observed kinetics of $(n\text{-Bu})_3\text{SnH}$ reduction of DCP and TCH. The agreement is good, and provides us with a knowledge of the E–V stereosequence as a function of comonomer composition.

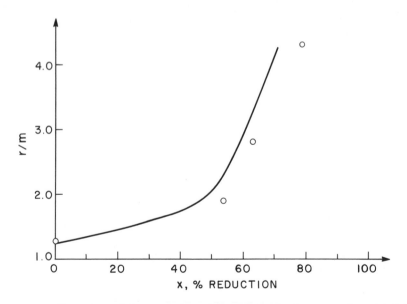

Figure 9.11 ■ Comparison of the observed (symbols) and simulated (curve) ratios r/m for VV diads during the $(n\text{-Bu})_3\text{SnH}$ reduction of PVC. [Reprinted with permission from Jameison et al. (1986, 1988).]

The excellent agreement between the simulated and observed reduction of PVC with $(n\text{-Bu})_3\text{SnH}$ means that both DCP and TCH are appropriate model compounds for the study of PVC reduction. DCP is useful for obtaining kinetic information on the relative reactivities of m and r VV diads and VV and EV diads. Reduction of TCH yields relative reactivities of the central and terminal chlorines in the VVV triads.

The physical properties of these E–V copolymers generated by the $(n\text{-Bu})_3\text{SnH}$ reduction of PVC have been demonstrated to be sensitive to their detailed microstructures, including their comonomer and stereosequence distributions (Tonelli et al., 1983; Bowmer and Tonelli, 1985, 1986a, 1986b, 1987; Tonelli and Valenciano, 1986; Gomez et al., 1987, 1989). ^{13}C NMR analysis of E–V copolymers (Schilling et al., 1985) yields microstructural information up to and including the comonomer triad level. However, properties such as crystallinity depend on E–V microstructure on a scale larger than comonomer triads. For example, the amount and stability of the crystals formed in E–V copolymers depend on the numbers and lengths of uninterrupted all-E-unit runs. The ability to computer-simulate the $(n\text{-Bu})_3\text{SnH}$ reduction of PVC makes available this information concerning the longer comonomer sequences in the resultant E–V copolymers.

Figure 9.12 presents the percentage of E units that are found in uninterrupted, all-E-unit runs as a function of the length of each run for two levels of reduction (70 and 90%). These data were obtained in two ways: (i) simulation

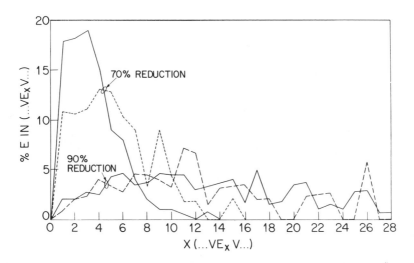

Figure 9.12 ■ Percentage of E units in uninterrupted all-E-unit runs as a function of run length and degree of reduction. Solid lines correspond to the computer simulation of PVC reduction with $(n\text{-Bu})_3\text{SnH}$, and dashed lines to the simulated results obtained by assuming random chlorine removal. [Reprinted with permission from Jameison et al. (1988).]

of the $(n\text{-Bu})_3\text{SnH}$ reduction of PVC and (ii) assuming random removal of chlorine during the reduction. The simulated data in Figure 9.12 make clear that the numbers and lengths of all-E-unit runs in E–V copolymers obtained via the reduction of PVC with $(n\text{-Bu})_3\text{SnH}$ are significantly reduced from those resulting from the random removal of chlorine. Though not shown in Figure 9.12, the percentage of E units in all-E-unit runs $\ldots\text{VE}_x\text{V}\ldots$ with $x > 29$ is 27% for random chlorine removal and just 14% for the $(n\text{-Bu})_3\text{SnH}$-reduced PVC. This observation has to be considered, for example, when discussing the crystalline morphology of E–V copolymers obtained by reduction of PVC with $(n\text{-Bu})_3\text{SnH}$ (Gomez et al., 1987, 1989).

9.3. Modification of 1,4-Poly(butadienes) with Dihalocarbenes

Because of the reactivity of its double bonds, 1,4-polybutadiene (PBD) may be readily modified to form a wide variety of homo- and copolymers (Pinazzi et al., 1969, 1975; Pinazzi and Levesque, 1967; Schilling et al., 1983). An interesting example, which has received close examination by a variety of NMR techniques (Siddiqui and Cais, 1986a, b; Cais and Siddiqui, 1987; Cais et al., 1987), is the addition of a carbene species to yield a saturated backbone incorporating cyclopropyl moieties. Cais and coworkers have reacted *cis*- and *trans*-PBD (*c*-PBD, *t*-PBD) with dichloro- and difluoro-, and fluorochloro-carbenes and studied the microstructures of their adducts with ^1H, ^{13}C, and ^{19}F NMR using both one- and two-dimensional techniques.

9.3.1. Possible Microstructures in the Dihalocarbene Adducts of PBD

As expected for singlet carbenes (Kirmse, 1971), cyclopropanation of the double bonds is stereospecific, i.e., the *trans* or *cis* character of the double bonds is preserved during the formation of the dihalocarbene adducts with PBD. We adopt the notation D and C to represent unreacted

$$\text{+CH}_2\text{—CH}\text{=}\text{CH—CH}_2\text{+}$$

and reacted

$$\begin{array}{c} \text{X} \diagdown \quad \diagup \text{Y} \\ \text{C} \\ \diagup \quad \diagdown \\ (\text{—CH}_2\text{—CH—CH—CH}_2\text{—}) \end{array}$$

units, respectively, where X, Y = F or Cl. Aside from the overall degree of conversion (mol% C), to completely characterize the microstructures of copolymers generated by dihalocarbene addition to PBD we must be able to specify their comonomer sequence distributions, the stereosequences of adjacent C units, and for the fluorochloro adducts, the *syn* and *anti* arrangements of adjacent cyclopropyl units in *c*-PBD : CFCl.

At the comonomer-triad level, we must be able to identify, or distinguish among, the six microstructures illustrated below for the dihalocarbene adduct with c-PBD.

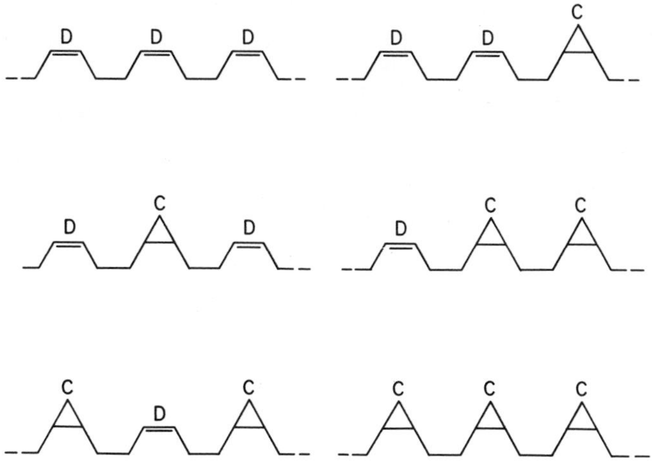

For microstructures containing adjacent C units, the stereochemistry of the dihalocarbene addition enters as an additional feature as indicated in the following difluorocarbene C–C diads:

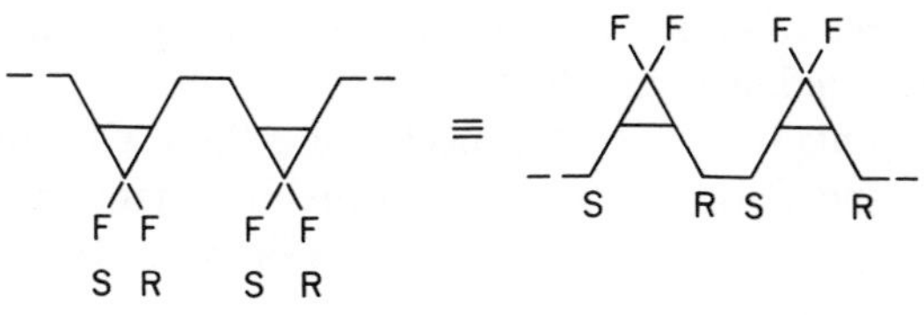

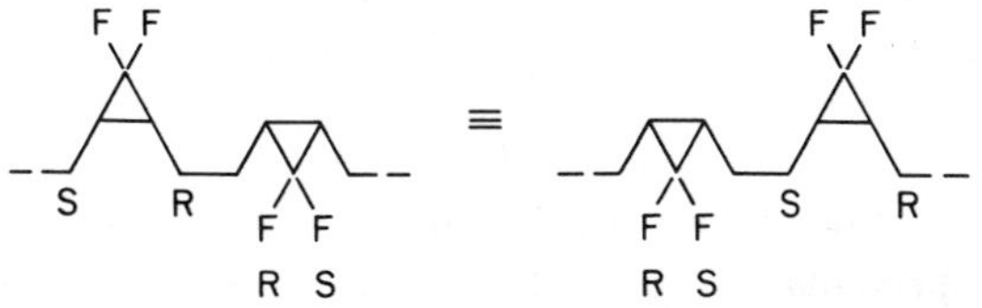

Finally, for unsymmetrical dihalocarbene adducts with c-PBD, there exist several relative orientations for isolated and neighboring cyclopropyl units. This type of isomerism is analogous to regiosequence isomers (see Chapter 7), and several examples are illustrated below. Note that the F and Cl atoms in the t-PBD : CFCl adducts are inchained in only a single orientation:

SYN-CI   ANTI-CI   DC

SYN-CI ANTI-CI   CC

CIS   TRANS

9.3.2. NMR of Dihalocarbene Adducts of PBD

The 500-MHz ^{1}H NMR spectra of c- and t-PBD : CF$_2$ adducts do not exhibit comonomer sequence fine structure. However, in the ^{1}H NMR spectra of both *cis* and *trans* adducts (see Figure 9.13) the allylic and olefinic protons from unreacted double bonds in D units are well separated from the remaining C-unit protons. Consequently, ^{1}H NMR provides the simplest means of measuring the degree of conversion (mol% C) via integration of appropriate peak areas, as indicated in Figure 9.13.

Figure 9.14 presents the ^{13}C NMR spectra of c-PBD : CF$_2$ and t-PBD : CF$_2$ adducts. Siddiqui and Cais (1986a) assigned the protonated carbons by spectral editing with the DEPT pulse sequence (Derome, 1987). The olefinic regions of the ^{13}C NMR spectra are found to be most sensitive to comonomer sequences. Figures 9.15 and 9.16 present the olefinic regions of the ^{13}C NMR spectra of several c-PBD : CF$_2$ and t-PBD : CF$_2$ adducts. Detailed assignments of this region of the spectra are presented in Tables 9.5 and 9.6, where a sensitivity to pentad comonomer sequences is revealed. Siddiqui and Cais (1986a) found that assumption of a random, or Bernoullian, distribution of comonomer sequences permitted a faithful simulation of the observed olefinic regions of the ^{13}C NMR spectra for both the c- and t-PBD adducts with :CF$_2$. Clearly the cyclopropanation of double bonds in c- and t-PBD by difluoro-

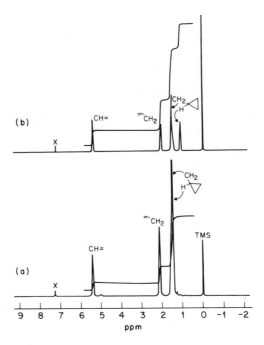

Figure 9.13 ■ 500-MHz ^{1}H NMR of (a) *cis*-PBD : CF$_2$ and (b) *trans*-PBD : CF$_2$ adducts, observed in CDCl$_3$ at room temperature. x = CHCl$_3$. [Reprinted with permission from Siddiqui and Cais (1986a).]

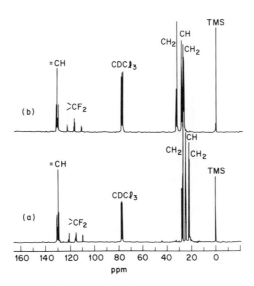

Figure 9.14 ■ 50.3-MHz ^{13}C NMR spectra of (a) *cis*-PBD : CF$_2$ and (b) *trans*-PBD : CF$_2$ adducts observed in CDCl$_3$ at 50°C. [Reprinted with permission from Siddiqui and Cais (1986a).]

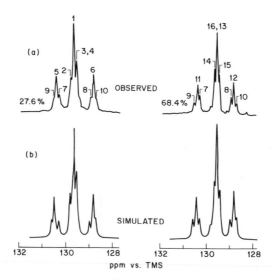

Figure 9.15 ■ (a) Olefinic-carbon region of 50.3-MHz ^{13}C spectra of *cis*-PBD : CF$_2$ adducts at indicated conversions, observed in CDCl$_3$ at 50°C. (b) Computer simulation of the ^{13}C olefinic region of the *cis* adducts for the corresponding conversions. The simulated spectra were generated by using the appropriate chemical shifts and constant line widths (4 Hz); Bernoullian statistics was assumed for the simulation. [Reprinted with permission from Siddiqui and Cais (1986a).]

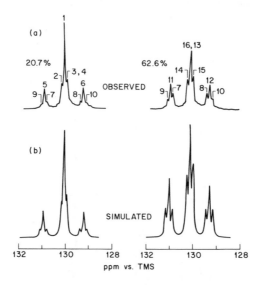

Figure 9.16 ■ (a) Olefinic carbon region of 50.3-MHz ^{13}C spectra of *trans*-PBD : CF$_2$ adducts at indicated conversions, observed in CDCl$_3$ at 50°C. (b) Computer simulation of the ^{13}C olefinic region of the *trans* adducts for the corresponding conversions. The simulated spectra were generated by using the appropriate chemical shifts and constant linewidths (4 Hz); Bernoullian statistics was assumed for the simulation. [Reprinted with permission from Siddiqui and Cais (1986a).]

Table 9.5 ▪ ^{13}C Chemical Shifts of Olefinic Carbons of *cis*-PBD : CF$_2$ Copolymers[a]

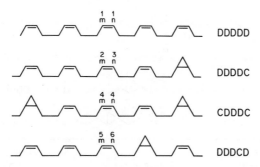

Peak Assignment	C-type sequence	Chem. shift vs. Me$_4$Si, ppm
1	DDDDD-*m, n*	129.68
2	DDDDC-*m*	129.79
3	DDDDC-*n*	129.54
4	CDDDC-*m, n*	129.54
5	DDDCD-*m*	130.41
6	DDDCD-*n*	128.81
7	CDDCD-*m*	130.27
8	CDDCD-*n*	128.92
9	DDDCC-*m*	130.53
10	DDDCC-*n*	128.69
11	CDDCC-*m*	130.38
12	CDDCC-*n*	128.84
13	DCDCD-*m, n*	129.55
14	DCDCC-*m*	129.66
15	DCDCC-*n*	129.45
16	CCDCC-*m, n*	129.55

[a]See Figure 9.15.
[b]Structural formula of some representative sequences are shown.

carbene is a random process independent of the comonomer sequence containing the reacting double bond.

The aliphatic regions of the ^{13}C NMR spectra of *c*-PBD : CF$_2$ and *t*-PBD : CF$_2$ are presented in Figures 9.17 and 9.18. The detailed assignments of these spectra, as given by Siddiqui and Cais (1986a), are presented in Tables 9.7 and 9.8. In addition to the sensitivity of the aliphatic CH$_2$ and CH carbons to comonomer triad sequences, both carbons are also sensitive to the stereosequences of adjacent CC diads and CCC triads. Siddiqui and Cais (1986a) found a random distribution of stereosequences in both the *c*-PBD : CF$_2$ and *t*-PBD : CF$_2$ adducts.

The 470.7-MHz ^{19}F NMR spectra of the asymmetric *c*-PBD adducts with fluorochlorocarbene are presented in Figure 9.19. The two widely spaced

Table 9.6 ▪ ^{13}C Chemical Shifts of Olefinic Carbons of *trans*-PBD : CF_2[a]

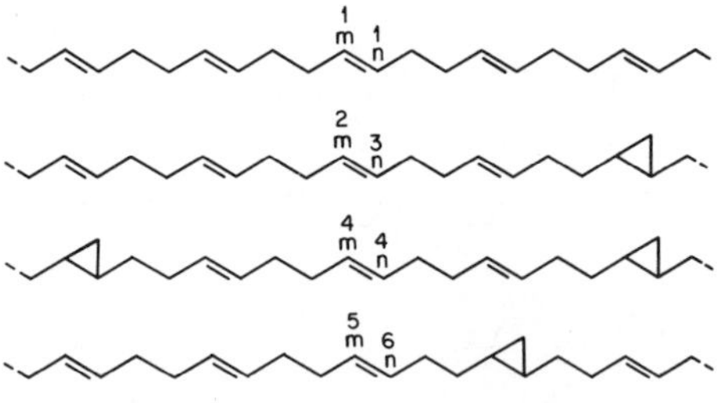

		DDDDD
		DDDDC
		CDDDC
		DDDCD

Peak Assignment	C-type sequence	Chem. shift vs. Me_4Si, ppm
1	DDDDD-m, n	130.06
2	DDDDC-m	130.18
3	DDDDC-n	129.91
4	CDDDC-m, n	129.91
5	DDDCD-m	130.94
6	DDDCD-n	129.24
7	CDDCD-m	130.81
8	CDDCD-n	129.35
9	DDDCC-m	131.03
10	DDDCC-n	129.10
11	CDDCC-m	130.93
12	CDDCC-n	129.23
13	DCDCD-m, n	130.09
14	DCDCC-m	130.20
15	DCDCC-n	129.96
16	CCDCC-m, n	130.11

[a]See Figure 9.16.
[b]Structural formulas of some representative sequences are shown.

groups of signals at -125.8 and -163.7 ppm correspond to the *syn*-Cl and *anti*-Cl structures shown below:

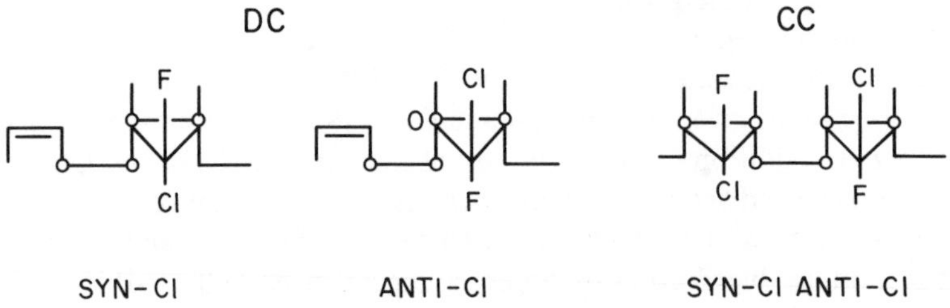

DC   CC

SYN–Cl   ANTI–Cl   SYN–Cl ANTI–Cl

These assignments are confirmed by a 2D heterocorrelated NOESY experiment [Cais et al. (1987) and see Chapter 8] and are consistent with the

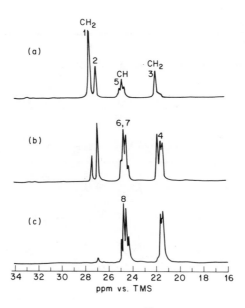

Figure 9.17 ■ Aliphatic-carbon region of 50.3-MHz ^{13}C spectra of *cis*-PBD : CF$_2$ adducts at (a) 27.6%, (b) 68.4%, and (c) 97.4% double-bond conversions, observed in CDCl$_3$ at 50°C. [Reprinted with permission from Siddiqui and Cais (1986a).]

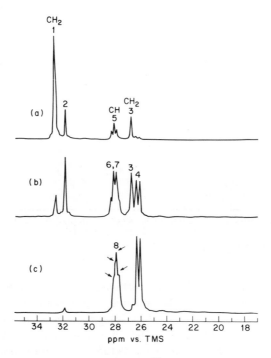

Figure 9.18 ■ Aliphatic-carbon region of 50.3-MHz ^{13}C spectra of *trans*-PBD : CF$_2$ adducts at (a) 20.7%, (b) 62.6%, and (c) 98.0% double-bond conversions, observed in CDCl$_3$ at 50°C. Arrows in (c) indicate four poorly resolved lines. [Reprinted with permission from Siddiqui and Cais (1986a).]

Table 9.7 ▪ ^{13}C Chemical Shifts of Aliphatic Carbons of *cis*-PBD : CF$_2$ Adducts^a

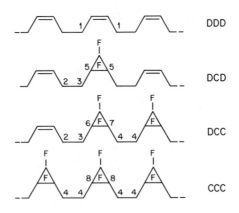

Peak Assignment	Sequence	Chem. shift vs. Me$_4$Si, ppm
1	DD	27.51^b
		27.45
2	DC, CD	26.94
3	DC, CD	21.87
4	CC	21.51
	(*m* and *r*)	21.33
5	DCD	24.92
	(mainly triplet)	24.72
		24.52
6	DCC (*m* and *r*)	24.91
7	CCD (*m′* and *r′*)	24.87
		24.71
		24.50
8	CCC	24.84
	(*rr′*, *rm′*, *mm′*, and *mr′*)	24.63
		24.46
		24.27

^{a}See Figure 9.17. Structural formulas for sequence determinations are shown.
^{b}The fine splittings may arise from the presence of either higher-order monomer sequences or long-range ^{13}C–^{19}F coupling.

shielding of fluorine nuclei by two methylene carbon γ-substituents in the *anti*-Cl structure, compared to the two hydrogen γ-substituents in the *syn*-Cl structure (see Chapter 7, Section 7.2.2). Integrated peak intensities indicate a *syn* : *anti* ratio of 1.7 : 1.0 (63% *syn*), which is independent of conversion.

The low-field signal due to the *syn* structure is split into contributions from DCD, DCC, and CCC comonomer sequence triads, with the CCC sequence further split into three components. Because the analogous fluorines in

Table 9.8 ■ ^{13}C Chemical Shifts of Aliphatic Carbons of *trans*-PBD : CF_2 Adducts[a]

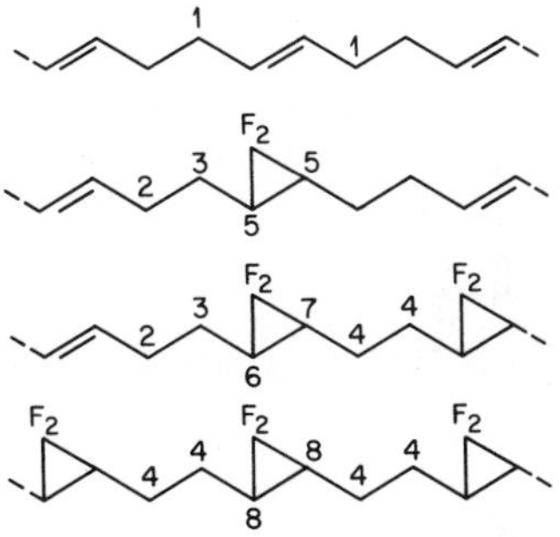

DDD

DCD

DCC

CCC

Peak Assignment	Sequence	Chem. shifts vs. Me_4Si, ppm
1	DD	32.70
2	CD, DC	31.88
2	CD, DC	31.88
3	CD, DC	26.80
4	CC	26.33
	(*m* and *r*)	26.06
5	DCD	28.27
	(mainly triplet)	28.08
		27.89
6	DCC (*m* and *r*)	28.30
7	CCD (*m′* and *r′*)	28.04
		27.89
		27.75
8	CCC	28.13
	(*rr′*, *rm′*, *mm′*, and *mr′*)	27.95
		27.89
		27.72

[a]See Figure 9.18. Structural formulas for sequence assignments are shown.

PBD : CF_2 adducts are insensitive to stereosequence (Siddiqui and Cais, 1986a), the CCC fine structure is probably due to *syn–syn–syn*, *syn–syn–anti* (*anti–syn–syn*), and *anti–syn–anti* arrangements. If these follow a Bernoullian distribution, their intensity ratios should be 2.8 : 3.2 : 1.0, respectively, based on a *syn* probability of 0.63. The actual observed ratios are 5.0 : 11.0 : 1.0. It appears that a neighboring-group effect is operative in this case, leading to a non-Bernoullian distribution of *syn* and *anti* placements in *c*-PBD : CFCl.

Figure 9.20 presents the 470.7-MHz ^{19}F NMR spectra of several *t*-PBD : CF–Cl adducts. The fluorine always has one hydrogen and one carbon γ-substituent in the *trans* enchained rings, so its chemical shift falls, as

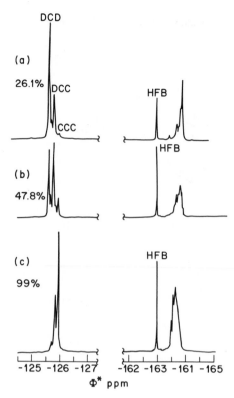

Figure 9.19 ■ 470.7-MHz ^{19}F NMR spectra from *cis*-PBD : CFCl having 26.1% (a), 47.8% (b), and 99% (c) conversion. The chemical-shift reference is hexafluorobenzene (HFB) at -163 ppm. The downfield signals arise from the *syn* isomer, and the upfield signals arise from the *anti* isomer. [Reprinted with permission from Cais and Siddiqui (1987).]

expected, in the middle of the widely spaced doublet seen for *c*-PBD : CFCl:

Though there is no *syn–anti* distribution in this case, the ^{19}F resonance is a complicated multiplet reflecting structural isomerism.

At low conversion we observe the effects of comonomer sequence triads. In those comonomer sequences with neighboring C units (DCC and CCC), further splitting is observed owing to different stereosequences. Cais and Siddiqui (1987) were able to conclude that *t*-PBD : CFCl is highly stereoirregular [see Figure 9.20(c)].

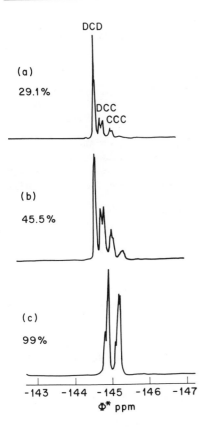

Figure 9.20 ■ 470.7-MHz ^{19}F NMR spectra from *trans*-PBD:CFCl having 29.1% (a). 45.5% (b), and 99% (c) conversion. [Reprinted with permission from Cais and Siddiqui (1987).]

By application of ^{1}H, ^{13}C, and ^{19}F NMR spectroscopy, in both the 1D and 2D modes, Cais and coworkers (Siddiqui and Cais, 1986a,b; Cais and Siddiqui, 1987; Cais et al., 1987) were able to almost completely unravel the microstructures of the dihalocarbene adducts with *c*- and *t*-PBD. Comonomer sequences at the pentad level, stereosequences at the CCC triad level, and (in *c*-PBD:CFCl) *syn–anti* isomerism also at the CCC triad level were identified. Armed with this wealth of microstructural, detail they have begun to investigate the physical properties of *c*- and *t*-PBD:CXY in an attempt to establish their structure–property relations.

References

Benson, S. (1960). *The Foundations of Chemical Kinetics*, McGraw-Hill, New York.

Bowmer, T. N. and Tonelli, A. E. (1985). *Polymer (British)* **26**, 1195.

Bowmer, T. N. and Tonelli, A. E. (1986a). *Macromolecules* **19**, 498.

Bowmer, T. N. and Tonelli, A. E. (1986b). *J. Polym. Sci. Polym. Phys. Ed.* **24**, 1631.

Bowmer, T. N. and Tonelli, A. E. (1987). *J. Polym. Sci. Polym. Phys. Ed.* **25**, 1153.

Cais, R. E. and Siddiqui, S. (1987). *Macromolecules* **20**, 1004.

Cais, R. E., Mirau, P. A., and Siddiqui, S. (1987). *British Polym. J.* **19**, 189.

Derome, A. E. (1987). *Modern NMR Techniques for Chemistry Research*, Pergamon, New York, Chapter 4.

Freeman, R. and Hill, H. D. W. (1971). *J. Chem. Phys.* **54**, 3367.

Gomez, M. A., Cozine, M. H., Tonelli, A. E., Schilling, F. C., Lovinger, A. J., and Davis, D. D. (1987). *Bull. Am. Phys. Soc.* **32** (3), 740.

Gomez, M. A., Cozine, M. H., Tonelli, A. E., Schilling, F. C., Lovinger, A. J., and Davis, D. D. (1989). *Macromolecules* **22**, in press.

Hagiwara, M., Miura, T., and Kagiya, T. (1969). *J. Polym. Sci. Part A-1* **7**, 513.

Jameison, F. A., Schilling, F. C., and Tonelli, A. E. (1986). *Macromolecules* **19**, 2168.

Jameison, F. A., Schilling, F. C., and Tonelli, A. E. (1988). *Chemical Reactions on Polymers*, ACS Symposium Series No. 364, J. L. Benham and J. F. Kinstle, Eds., Am. Chem. Soc., Washington, p. 356.

Keller, F. and Mugge, C. (1976). *Faserforsch. Textiltech.* **27**, 347.

Kirmse, W. (1971). *Carbene Chemistry*, Second Ed., Academic Press, New York.

Mark, J. E. (1973). *Polymer (British)* **14**, 553.

Misono, A., Uchida, Y., and Yamada, K. (1967). *J. Polym. Sci. Part B* **5**, 401; *Bull. Chem. Soc. Jpn.* **40**, 2366.

Misono, A., Uchida, Y., Yamada, K., and Saeki, T. (1968). *Bull. Chem. Soc. Jpn.* **41**, 2995.

Pinazzi, C. and Levesque, G. (1967). *C. R. Acad. Sci. Paris Ser. C* **264**, 288.

Pinazzi, C., Gueniffey, H., Levesque, G., Reyx, D., and Pleurdeau, A. (1969). *J. Polym. Sci. Part C* **22**, 1161.

Pinazzi, C., Brosse, J. C., Pleurdeau, A., and Reyx, D. (1975). *Appl. Polym. Symp.* **26**, 73.

Schilling, F. C., Bovey, F. A., Tseng, S., and Woodward, A. E. (1983). *Macromolecules* **16**, 808.

Schilling, F. C., Tonelli, A. E., and Valenciano, M. (1985). *Macromolecules* **18**, 356.

Siddiqui, S. and Cais, R. E. (1986a). *Macromolecules* **19**, 595.

Siddiqui, S. and Cais, R. E. (1986b). *Macromolecules* **19**, 998.

Starnes, Jr., W. H., Plitz, I. M., Hische, D. C., Freed, D. J., Schilling, F. C., and Schilling, M. L. (1978). *Macromolecules* **11**, 373.

Starnes, Jr., W. H., Schilling, F. C., Abbas, K., Plitz, I. M., Hartless, R. L., and Bovey, F. A. (1979). *Macromolecules* **12**, 13.

Starnes, Jr., W. H., Schilling, F. C., Plitz, I. M., Cais, R. E., Freed, D. J., Hartless, R. L., and Bovey, F. A. (1983). *Macromolecules* **16**, 790 and references cited therein.

Stothers, J. B. (1972). *Carbon-13 NMR Spectroscopy*, Academic Press, New York, Chapter 5.

Tonelli, A. E. and Schilling, F. C. (1981). *Macromolecules* **14**, 74.

Tonelli, A. E., Schilling, F. C., Starnes, Jr., W. H., Shepherd, L., and Plitz, I. M. (1979). *Macromolecules* **12**, 78.

Tonelli, A. E., Schilling, F. C., Bowmer, T. N., and Valenciano, M. (1983). *Polym. Preprints Am. Chem. Soc. Div. Polym. Chem.* **24** (2), 211.

Tonelli, A. E. and Valenciano, M. (1986). *Macromolecules* **19**, 2643.

10

Biopolymers

10.1. Introduction

The NMR spectra of biopolymers are generally very complex, because they reflect the molecular complexity of biopolymer microstructures (MacGregor and Greenwood, 1980). This microstructural complexity is evident in Figure 10.1, where a portion of a polypeptide, or protein, chain is illustrated. Each monomer unit, or residue, in a polypeptide chain can possess any of the 20 or so different side-group substituents R corresponding to the naturally occurring amino and imino acids listed in Table 10.1. As a consequence, polypeptides and proteins are really copolymers consisting of some 20 unique comonomers enchained in a myriad of possible comonomer sequences. Elucidation of their microstructures requires establishment of the order in which each of the approximately 20 possible amino and imino acid residues is incorporated into their chains. An example is provided by the protein hormone insulin shown in Figure 10.2.

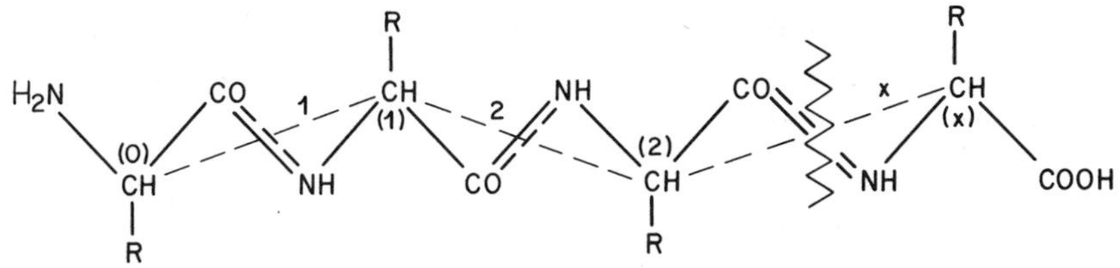

Figure 10.1 ■ A polypeptide chain. [Reprinted with permission from Flory (1969).] The partial double bond character of the amide or peptide bonds is indicated. The sequential numbering of residues is also shown.

Table 10.1 ■ The Common Amino and Imino Acid Residues Found in Proteins

Name	Symbol	Structure	Name	Symbol	Structure
Alanine	Ala	CH_3 $\mid$ $-NHCH_\alpha CO-$	Glutamine	Gln	$CONH_2$ $\mid$ $CH_{2\gamma}$ $\mid$ $CH_{2\beta}$ $\mid$ $-NHCH_\alpha CO-$
Arginine	Arg	$HN=C\overset{\displaystyle NH_2}{\underset{\displaystyle NH}{\big<}}$ $\mid$ $CH_{2\delta}$ $\mid$ $CH_{2\gamma}$ $\mid$ $CH_{2\beta}$ $\mid$ $-NHCH_\alpha CO-$	Glycine	Gly	$-NHCH_{2\alpha}CO-$
			Hydroxy-proline	Hyp	OH $\langle\delta^{\,\gamma}\beta\rangle$ $-NH-CH_\alpha CO-$
Asparagine	Asn	$CONH_2$ $\mid$ $CH_{2\beta}$ $\mid$ $-NHCH_\alpha CO-$	Histidine	His	$CH_{2\beta}$ $\mid$ $-NHCH_\alpha CO-$
Aspartic acid	Asp	CO_2H $\mid$ $CH_{2\beta}$ $\mid$ $-NHCH_\alpha CO-$	Isoleucine	Ile	$CH_{3\delta}\quad CH_{2\gamma}CH_{3\delta}$ CH_β $\mid$ $-NHCH_\alpha CO-$
Cysteine	Cys	SH $\mid$ $CH_{2\beta}$ $\mid$ $-NHCH_\alpha CO-$	Leucine	Leu	$CH_{3\delta'}\quad CH_{3\delta}$ CH_γ $\mid$ $CH_{2\beta}$ $\mid$ $-NHCH_\alpha CO-$
Cystine	Cys	$S\tfrac{}{}_2$ $\mid$ $CH_{2\beta}$ $\mid$ $-NHCH_\alpha CO-$	Lysine	Lys	$CH_{2\gamma}CH_{2\delta}CH_{2\epsilon}NH_2$ $\mid$ $CH_{2\beta}$ $\mid$ $-NHCH_\alpha CO-$
Glutamic acid	Glu	CO_2H $\mid$ $CH_{2\gamma}$ $\mid$ $CH_{2\beta}$ $\mid$ $-NHCH_\alpha CO-$	Methionine	Met	$CH_{3\epsilon}$ $\mid$ S $\mid$ $CH_{2\gamma}$ $\mid$ $CH_{2\beta}$ $\mid$ $-NHCH_\alpha CO-$

Table 10.1 ■

Name	Symbol	Structure	Name	Symbol	Structure
Ornithine	Orn	$CH_{2\gamma}CH_{2\delta}NH_2$ $\overset{\mid}{CH_{2\beta}}$ $\overset{\mid}{-NHCH_{2\alpha}CO-}$	Tryptophan	Trp	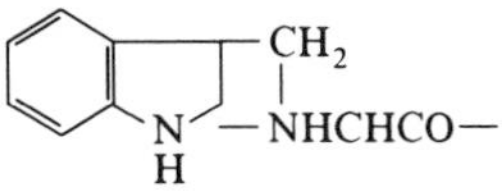
Phenylalanine	Phe	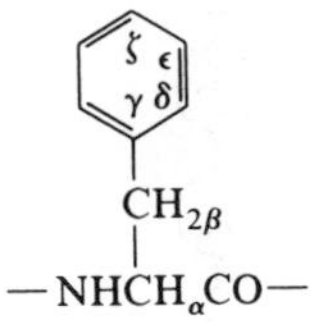	Tyrosine	Tyr	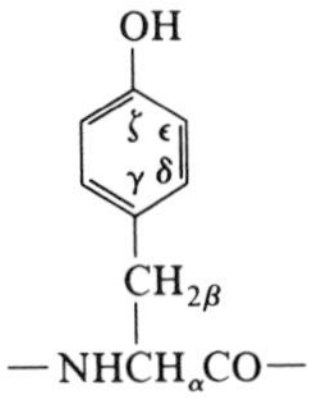
Proline	Pro	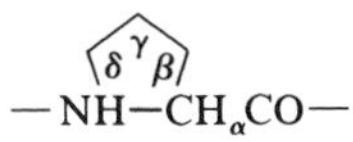	Valine	Val	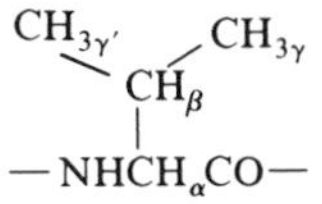 $CH_{3\gamma'}\diagdown\,\diagup CH_{3\gamma}$ CH_β $\overset{\mid}{-NHCH_\alpha CO-}$
Serine	Ser	$CH_{2\beta}OH$ $\overset{\mid}{-NHCH_\alpha CO-}$			
Threonine	Thr	$CH_{3\gamma}$ $\overset{\mid}{CH_\beta OH}$ $\overset{\mid}{-NHCH_\alpha CO-}$			

Before NMR spectroscopy can be successfully utilized to assist in the determination of biopolymer microstructure, we must be able to assign the observed resonances to the individual monomer units which constitute the biopolymer. Let us illustrate the complexity of this task by way of the synthetic cyclic polypeptide drawn schematically in Figure 10.3. [Note that the Ala residues (see Table 10.1) are both of the D configuration in this model polypeptide, while virtually all of the amino and imino acid residues in proteins are of the L configuration.]

The 220-MHz ^{1}H NMR spectrum of 4 Gly-2 D-Ala recorded in DMSO-d_6 at 25°C (Tonelli and Brewster, 1972) is presented in Figure 10.4, where only the amide proton (NH) region is shown. Even though this simple model polypeptide contains only two chemically distinct amino acid residues (Gly and D-Ala), each of the six NH protons exhibits a unique resonance multiplet. The D-Ala doublets at 8.06 and 8.17 ppm and the three triplets and one pair of doublets (7.89 ppm) from the Gly residues, which originate from the

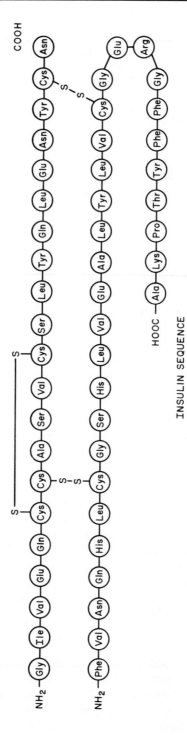

INSULIN SEQUENCE

Figure 10.2 ■ The peptide residue sequence of the hormone insulin.

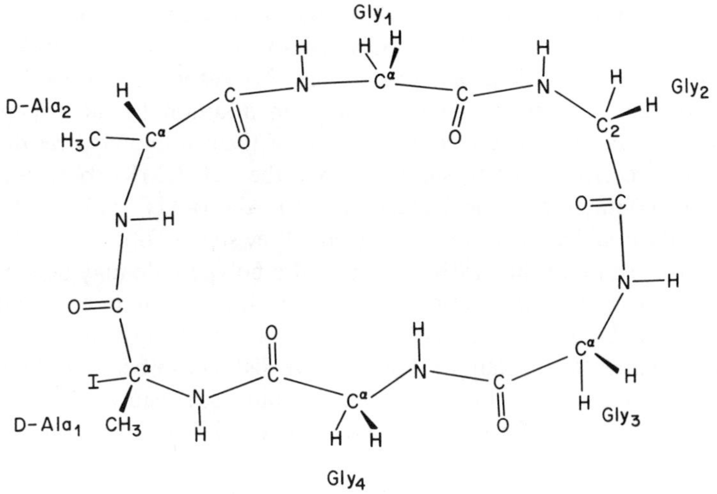

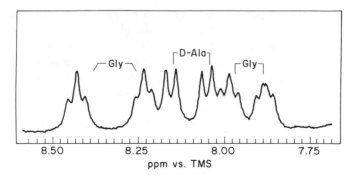

Figure 10.3 ▪ Schematic drawing of the chemical structure of the synthetic cyclic hexapeptide 4 Gly–2D-Ala.

vicinal, three-bond coupling of NH to $C^\alpha H$ protons, are all separately visible. Clearly the ^1H NMR spectrum of the deceptively simple cyclic hexapeptide 4 Gly–2 D-Ala is sensitive to its microstructure.

If the chemical shift of each NH proton depended solely on the chemical structure of its constituent amino acid residues, then we would expect to observe only two resonance multiplets: one for the four Gly residues and another for the two D-Ala residues. If the NH chemical shifts are dependent upon the longer-range microstructure characterized by their nearest-neighbor residue sequence, then we would expect to observe five distinct resonance

Figure 10.4 ▪ The NH region of the 220-MHz ^1H NMR spectrum of 4 Gly–2 D-Ala in DMSO-d at 25°C. [Reprinted with permission from Tonelli and Brewster (1972).]

multiplets corresponding to the five distinct tripeptide sequences –D-Ala–D-Ala–Gly–, –D-Ala–Gly–Gly–, –Gly–Gly–Gly–, –Gly–Gly–D-Ala–, and –Gly–D-Ala–D-Ala–. The fact that all six NH resonances are magnetically distinct means that their chemical shifts, in addition to the tripeptide sequences, are sensitive to the conformations of their amino acid residues. This conclusion receives further support from the ^{1}H NMR spectrum of the all-alanine-containing cyclic hexapeptide 5 L-Ala–D-Ala, which exhibits six distinct NH doublets (Tonelli and Richard Brewster, 1973).

We have seen that the NMR spectrum of a polypeptide may be sensitive to the sequence and conformations of its constituent amino and imino acid residues, but the question of assigning each observed resonance to a specific residue remains. As an example, how may we distinguish between the four Gly and three L-Ala residues in the hormone insulin (see Figure 10.2)? To utilize the microstructural information inherent in the NMR spectrum of a large polypeptide hormone like insulin, we must possess the means to identify and distinguish between the four Gly resonances corresponding to the NH—Gly—Ile—, —Cys—Gly—Ser—, —Cys—Gly—Glu—, and —Arg—Gly—Phe— sequences. Two-dimensional NMR techniques provide us with the means to assign complicated NMR spectra of polypeptides and small proteins (Wüthrich, 1986). J-correlated COSY spectra (see Chapter 6) establish the spin connectivities within and between the residues of a polypeptide chain, permitting assignment of resonances as to residue type and sequence.

In this chapter we describe several applications of 2D NMR to the study of the microstructures and conformations of biopolymers including polypeptides, polynucleotides, and polysaccharides.

10.2. Polypeptides

10.2.1. 2D NMR Assignment of ^{1}H Resonances

Glucagon is a linear polypeptide hormone of 29 amino acid residues (see Figure 10.5) that functions when bound to the plasma membranes of liver and other cells (Pohl et al., 1969). X-ray diffraction measurements performed on glucagon single crystals (Sasaki et al., 1975) yielded an α-helical conformation (see Section 10.2.2) in the solid, which is not preserved in aqueous solutions of monomeric glucagon according to ^{1}H NMR studies (Bosch et al., 1978). Wider et al. (1982) and Wüthrich et al. (1982) have studied, via 2D NMR, the conformation of glucagon bound to dodecylphosphocholine micelles in water. In this section we briefly indicate their assignment (Wider et al., 1982) of the ^{1}H NMR spectrum of glucagon.

The assignment strategy may be summarized in the following two steps: (i) 2D COSY spectra are used to establish intraresidue J-coupled spin connectivities, thereby establishing assignments of ^{1}H resonances to residue types (see

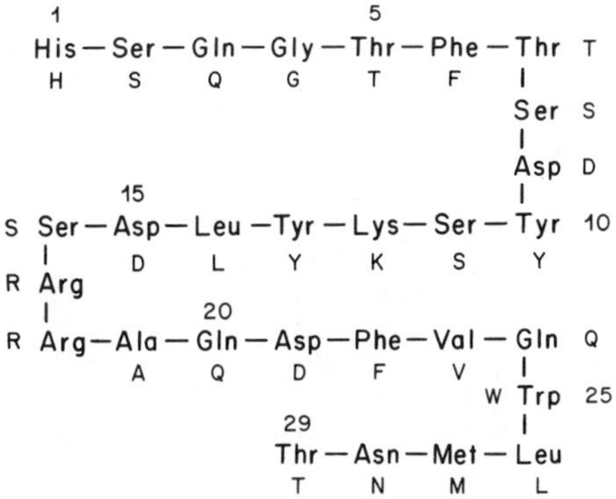

```
 1                      5
His — Ser — Gln — Gly — Thr — Phe — Thr  T
 H      S      Q      G      T      F     I
                                         Ser  S
                                          I
                                         Asp  D
            15                            I
 S  Ser — Asp — Leu — Tyr — Lys — Ser — Tyr  10
    I      D      L      Y      K      S     Y
 R Arg
    I            20
 R Arg — Ala — Gln — Asp — Phe — Val — Gln  Q
          A      Q      D      F      V     I
                                         W Trp  25
            29                            I
           Thr — Asn — Met — Leu
            T      N      M      L
```

Figure 10.5 ■ Primary structure of glucagon. Both the three- and the one-letter designations for the amino acid residues are indicated.

Figures 10.6, 10.7, and 10.8) and (ii) 2D ^{1}H NOESY spectra are used to determine amino acid residue connectivities via NOEs observed between NH_{i+1} and $C^\alpha H_i$, NH_{i+1} and NH_{i+2} or NH_i, and NH_{i+1} and $C^\beta H_i$, where Figure 10.9 illustrates the $NH_{i+1}-C^\beta H_i$ NOE connectivities. A schematic summary of the NOE-determined residue connectivities is presented in Figure 10.10.

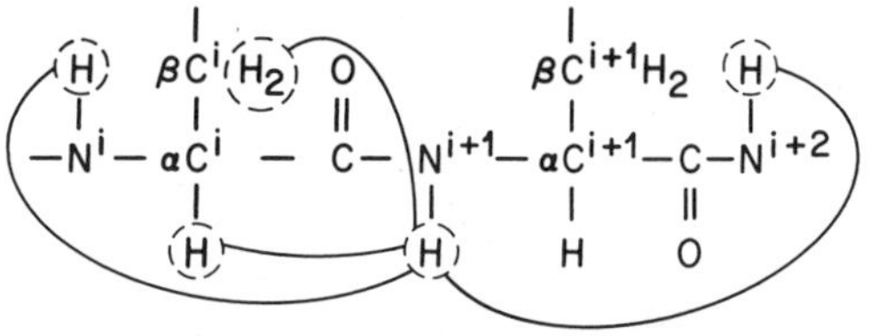

The complete assignment of ^{1}H NMR resonances observed for glucagon bound to dodecylphosphocholine micelles (Wider et al., 1982) is presented in Table 10.2. With these data in hand it is possible to study the conformation (secondary structure) of bound glucagon via interproton distances determined in 2D NOESY ^{1}H NMR experiments (Braun et al., 1983). A simpler example of the 2D NMR determination of polypeptide solution conformations is presented in the next section.

10.2.2. Determination of Polypeptide Conformation by 2D NMR

Here we present a brief discussion of the application of 2D NOESY ^{1}H NMR to determine the conformation of a polypeptide. The synthetic polypeptide

Table 10.2 ■ Chemical Shifts, δ^a, of the Assigned ^{1}H NMR Lines of Glucagon
Bound to Perdeuterated Dodecylphosphocholine Micelles[b]

Amino acid residue	$\delta\ (\pm 0.01\ \text{ppm})^a$			
	NH	$C^\alpha H$	$C^\beta H$	Others
His1[c]	n.o.	4.66[c]	3.07, 3.21[c]	$C^\delta H$ 7.22, $C^\epsilon H$ 8.11
Ser2[c]	n.o.	4.51[c]	3.86, 3.86[c]	
Gln3	8.75	4.41	2.01, 2.17	$C^\gamma H_2$, 2.38, 2.38
Gly4	8.47	4.02		
		4.02		
Thr5	8.06	4.34	4.19	$C^\gamma H_3$ 1.07
Phe6	8.63	4.64	3.13, 3.22	Ring 7.27
Thr7	8.11	4.23	4.24	$C^\gamma H_3$ 1.17
Ser8	8.04	4.37	3.78, 3.86	
Asp9	8.30	4.57	2.52, 2.52	
Tyr10	8.05	4.56	2.85, 3.18	$C^\delta H$ 7.06, 7.06
				$C^\epsilon H$ 6.82, 6.82
Ser11	8.03	4.06	3.96, 3.96	
Lys12	7.75	4.09	1.31, 1.59	$C^\gamma H_2$ 1.10, 1.52
				$C^\delta H_2$ 1.50, 1.50
				$C^\epsilon H_2$ 2.80, 2.80
Tyr13	7.55	4.42	2.89, 3.23	$C^\delta H$ 7.19, 7.19
				$C^\epsilon H$ 6.84, 6.84
Leu14	7.52	4.33	1.61, 1.81	$C^\gamma H$ 1.77
				$C^\delta H_3$ 0.88, 0.96
Asp15	7.55	4.75	2.63, 2.76	
Ser16	8.45	4.23	3.63, 3.63	
Arg17	8.50	4.17	1.96, 1.96	$C^\gamma H_2$ 1.73, 1.73
				$C^\delta H_2$ 3.25, 3.25
Arg18	8.24	4.25	1.86, 1.94	$C^\gamma H_2$ 1.71, 1.71
				$C^\delta H_2$ 3.27, 3.27
Ala19	7.99	4.17	1.56	
Gln20	8.17	4.02	2.25	
Asp21	8.48	4.47	2.62, 2.92	
Phe22	8.18	4.44	3.38, 3.38	Ring 7.23
Val23	8.28	3.49	2.26	$C^\gamma H_3$ 1.02, 1.21
Gln24	8.24	4.00	2.22	
Trp25	7.92	4.30	3.33, 3.62	$C^\delta H$ 7.37
				$N^\epsilon H$ 10.53
				$C^\epsilon H$ 7.33
				$C^{\zeta 2} H$ 7.54
				$C^{\zeta 3} H$ 6.89
				$C^\eta H$ 7.11
Leu26	8.09	3.27	1.41, 1.57	$C^\gamma H$ 1.56
				$C^\delta H_3$ 0.72, 0.72
Met27	7.79	4.30	2.03, 2.13	$C^\gamma H_2$ 2.53, 2.70
				$C^\epsilon H_3$ 2.04
Asn28	7.66	4.74	2.68, 2.97	
Thr29	7.53	4.04	4.11	$C^\gamma H_3$ 1.07

[a] The chemical shifts δ are relative to external sodium 3-trimethylsilyl-$[2,2,3,3-^2H_4]$propionate. The ϵ-methyl resonance of Met27 was taken to be at 2.04 ppm and was used as internal reference.
[b] pH 6.0, $T = 37°C$ (Wider et al., 1982).
[c] After sequential assignments for the amino acid residues 3 to 29, the remaining resonances were assigned to His1 and Ser2 based on comparison with the corresponding random-coil values.

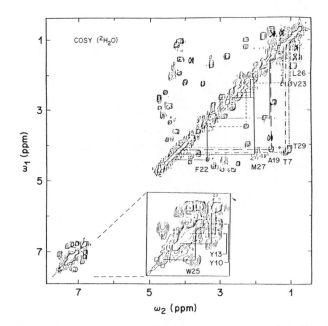

Figure 10.6 ■ Contour plot of a 360-MHz ^{1}H COSY spectrum of glucagon bound to perdeuterated dodecylphosphocholine micelles in ^{2}H$_2$O. The sample contained 0.015 M glucagon, 0.7 M [^{2}H$_{38}$]dodecylphosphocholine, 0.05 M phosphate buffer, p^2H 6.0, $t = 37°$C. Under these conditions the predominant species in the solution are mixed micelles of 1 glucagon molecule and ~ 40 detergent molecules, with a molecular weight of about 17,000. The spectrum was recorded in 24 h; the digital resolution is 5.88 Hz/point. The symmetrized absolute-value spectrum is shown. The aromatic region is also presented on an expanded scale. Proton–proton J-connectivities are indicated for the following residues: Thr7 (dot-dash lines), Ala19 (dashed), Phe22 (solid), Val23 (dashed), Leu26 (solid), Met27 (solid), Thr29 (dotted), and the aromatic rings of Tyr10 (dotted), Tyr13 (dashed), and Trp25 (solid). In order not to overcrowd the figure, only the C^αH connectivity with the lower-field C^βH lines is shown, even for amino acid residues where two nondegenerate β-methylene resonances were observed. Cross peaks originating from residual protons in the perdeuterated dodecylphosphocholine are marked $\times$. [Reprinted with permission from Wider et al. (1982).]

poly-γ-benzyl L-glutamate

$$\begin{array}{c} O \\ \parallel \\ C-O-CH_2 \cdot C_6H_5 \\ \mid \\ CH_{2\gamma} \qquad PBLG \\ \mid \\ CH_{2\beta} \quad O \\ \mid \qquad \parallel \\ \left[NH-CH - C \right]_n \\ \quad \alpha \end{array}$$

(PBLG) may exist in either α-helical or random coil form in solution, depend-

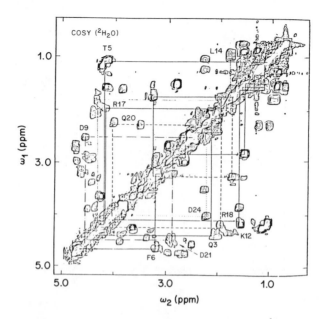

Figure 10.7 ▪ Spectral region from 0.2 to 5.1 ppm of the 360-MHz ^{1}H COSY spectrum of micelle-bound glucagon in Figure 10.6. *J*-connectivities for the following residues are indicated in the upper left triangle of the spectrum: Thr5 (solid lines), Asp9 (dot-dash), Leu14 (dotted), Arg17 (solid), Gln20 (dashed). The lower right triangle contains the *J*-connectivities for Gln3 (solid), Phe6 (solid), Lys12 (solid), Arg18 (dashed), Asp21 (dot-dash), Gln24 (dotted). In order not to overcrowd the figure, only the C$^\alpha$H connectivity with the lower-field C$^\beta$H line is shown, even for amino acid residues where two nondegenerate β-methylene resonances were observed. [Reprinted with permission from Wider et al. (1982).]

ing on the solvent and temperature. In Figure 10.11(a) a portion of a polypeptide chain is drawn, illustrating the bond rotation angles of the backbone (ϕ_i, ψ_i, ω_i) and the sidechain (χ_i^1, χ_i^2) (IUPAC-IUB, 1970) for a constituent amino acid residue. Below this drawing a Ramachandran steric map (Ramachandran et al., 1963) is presented, outlining the backbone conformations (ϕ, ψ), where the amide bonds are *trans* ($\omega = 180°$), that are allowed for a PBLG residue in the random coil form, i.e., all conformations (ϕ, ψ) within the dashed lines. The α-helical form corresponds to each amino acid residue being confined to the conformation (ϕ, ψ) = $-58°$, $-47°$ indicated by α on the steric map in Figure 10.11(b).

X-ray diffraction has demonstrated the α-helical form for crystalline PBLG, and the existence of the PBLG α-helix in solution is indicated by its optical properties. It is generally assumed that the solution and solid-state α-helices of PBLG are very similar. Recently Mirau and Bovey (1986) tested this assumption through measurement of interproton distances by 2D NOESY ^{1}H NMR (see Chapter 8).

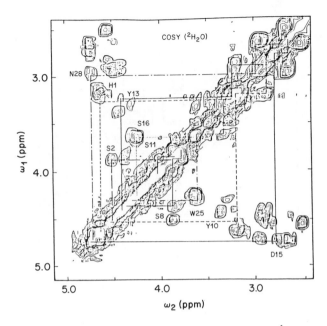

Figure 10.8 ■ Spectral region from 2.4 to 5.1 ppm of the 360-MHz ^{1}H COSY spectrum of micelle-bound glucagon in Figure 10.6. Connectivities for the following residues are indicated in the upper left triangle: His1 (dashed lines), Ser2 (dashed lines), Ser11 (solid), Tyr13 (solid), Ser16 (dotted), Asn28 (dot-dash). The lower right triangle contains the J-connectivities for Ser8 (solid), Tyr10 (dashed), Asp15 (solid), Trp25 (dot-dash). In order not to overcrowd the figure, only the C^αH connectivity with the lower-field C^βH line is shown, even for amino acid residues where two nondegenerate β-methylene resonances were observed. [Reprinted with permission from Wider et al. (1982).]

A portion of α-helical PBLG (14-mer) is drawn in Figure 10.12 along with the 500 MHz ^{1}H NMR spectrum of a PBLG 20-mer in α-helical form observed in CDCl$_3$ solution (Mirau and Bovey, 1986). We are principally interested in the NH, αH, and βH protons. The sidechains of PBLG are large enough to give the α-helical 20-mer an overall shape that is nearly as broad (20 Å) as it is long (30 Å). It is assumed that the PBLG 20-mers are tumbling isotropically in solution, because correlation times τ_c of 0.4 and 0.9 nsec are obtained for the tumbling of a prolate ellipsoid (Broersma, 1960) of these dimensions about the long and short axes, respectively. Additional 2D NOESY measurements (Mirau and Bovey, 1986) confirm this conclusion and yield an isotropic $\tau_c = 1$ nsec.

Figure 10.13(a) presents the 500-MHz 2D NOESY ^{1}H NMR spectrum of α-helical PBLG (Mirau and Bovey, 1986). Recall (Chapter 8) that in the NOESY experiment an additional 90° pulse is inserted between the two 90° pulses of the COSY experiment (see Figure 8.2). Following this additional

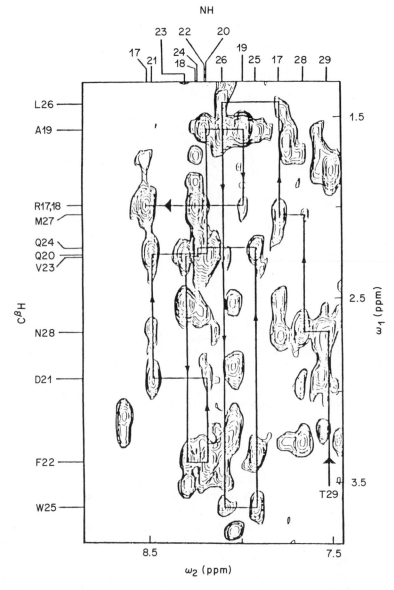

Figure 10.9 ■ Spectral region $(1.3–3.8 \text{ ppm}) \times (7.4–8.8 \text{ ppm})$ of the 360-MHz ^{1}H NOESY spectrum of micelle-bound glucagon. The continuous lines with arrows indicate the sequential assignments for the polypeptide segment residues 17 to 29, which were obtained from NOEs between amide protons and the C^β protons of the preceding residues. The numbers at the top of the figure indicate the amide proton chemical shifts of the corresponding residues; those on the left margin indicate the chemical shifts of one C^β proton for each residue. [Reprinted with permission from Wider et al. (1982).]

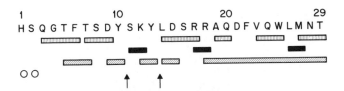

Figure 10.10 ▪ Amino acid sequence of bovine glucagon and survey of the experimental data by which individual resonance assignments were obtained for the micelle-bound polypeptide. Diagonal hatching: Sequential assignments via NOE from NH_{i+1} to $C^\alpha H_i$; solid: sequential assignments via NOE from NH_i to NH_{i+1}; vertical hatching: sequential assignments via NOE from NH_{i+1} to $C^\beta H_i$; ○: assignment in the sequence relied on the identification of the spin system in the COSY spectrum, whereby the amide proton resonances of these residues were not observed. The arrows indicate locations where all the resonances were assigned but the connectivity between two neighboring residues was not established. [Reprinted with permission from Wider et al. (1982).]

pulse there is a mixing time, τ_m, before the final pulse and acquisition of the spectrum. τ_m is on the order of T_1, and during τ_m some spins exchange magnetization by direct, through-space dipole–dipole interaction. Spins labeled by their frequencies during the evolution time t_1 after the first pulse may precess at a different frequency at the end of τ_m (after the second pulse), giving rise to cross peaks in the 2D NOESY spectrum. Measurement of cross-peak intensities as a function of τ_m permits a determination of the rates of magnetization transfer between ^1H spins, and from these transfer rates interproton distances may be calculated.

Figure 10.13(b) presents the growth of the cross peaks as a function of the mixing time τ_m. From the slopes of these plots the interproton distances r_{HH} are evaluated according to

$$\text{slope} = \sigma = 5.7\left[\frac{6\tau_c}{1 + 4\omega^2\tau_c^2} - \tau_c\right] \times 10^{10}r_{HH}^{-6}, \qquad (10.1)$$

where ω is the observing frequency (3.14×10^9 rad/sec) and τ_c the correlation time of the tumbling helix (10^{-9} sec). Table 10.3 presents three interproton distances calculated according to Eq. 10.1 for the α-helical PBLG 20-mer in solution, and they are compared with the same distances obtained from the x-ray structure of the crystalline α-helix. Though the distances are comparable, the backbone NH–αH distance differs by more than the experimental error (0.15 Å) and likely represents a real difference between the α-helical conformations of PBLG in the crystal and in solution.

10.3. Polynucleotides

Unlike polypeptides and proteins, which are built up from 20 or so different amino and imino acids, polynucleotides (DNAs) usually possess only four unique structural units: the purine nucleotides adenine (A) and guanine (G)

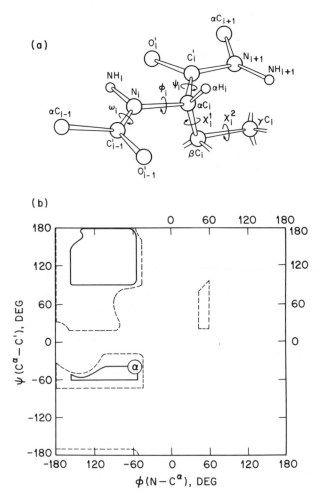

Figure 10.11 ▪ (a) Standard nomenclature for the atoms and torsion angles along a polypeptide chain (IUPAC-IUB, 1970). (b) Ramachandran steric map (Ramachandran et al., 1963) for the L-Ala residue, which is also appropriate for PBLG. "Normally allowed" random coil regions are enclosed by solid lines, and those encompassed by dashed lines correspond to "outer limit" interatomic distances. The α-helical conformation is indicated by α at $\phi = -58°$, $\psi = -47°$.

and the pyrimidine nucleotides cytosine (C) and thymine (T) (see Figure 10.14). Consequently, it is not difficult to assign resonances to the four different types of DNA residues, but the determination of what residue sequence and/or conformation corresponds to a given resonance remains a formidable task. 2D NMR techniques are the tools most useful in answering these microstructural and conformational questions for polynucleotides.

(a)

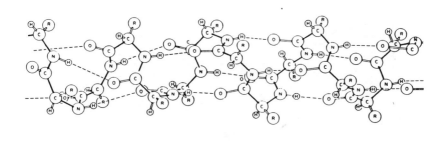

(b)

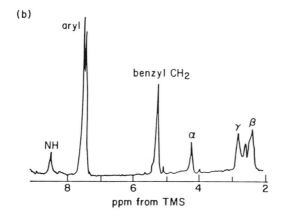

Figure 10.12 ■ (a) α-helical PBLG (14-mer). (b) 500-MHz ^{1}H NMR spectrum of α-helical PBLG in 95 : 5 chloroform : trifluoroacetic acid. The trifluoroacetic acid services to disrupt aggregates of PBLG α-helices, which themselves remain intact. [Reprinted with permission from Mirau and Bovey (1986).]

Figure 10.15 defines the torsion angles in a nucleotide which specify its conformation. According to x-ray diffraction studies of small nucleotides (Altona and Sundaralingam, 1973; Altona, 1975; Altona et al., 1976; Haasnoot et al., 1981) the two principal sugar-ring puckers are the 2'-endo and 3'-endo conformations shown in Figure 10.16. Also illustrated there are the different conformations obtained by rotations χ about the glycosidic bond connecting sugar and base rings. $\chi = 180 \pm 90°$ is referred to as the *anti* conformation and $\chi = 0 \pm 90°$ as the *syn* conformation. In the commonly observed right-handed DNA duplex structures A-DNA and B-DNA (see Figure 10.17) (Dickerson, 1983), all nucleotides assume the *anti* conformation, while in left-handed Z-DNA the nucleotide residues alternate between the *anti* and *syn* conformations. In A-DNA the sugar pucker is 3'-endo, while in

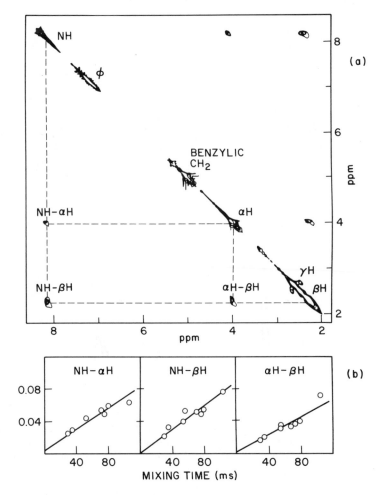

Figure 10.13 ■ (a) Absorption phase 2D NOE spectrum of α-helical PBLG 20-mer in 95 : 5 chloroform : trifluoroacetic acid. (b) Initial rise of cross-peak intensities in spectrum (a) as a function of mixing time τ_m; spectrum in (a) corresponds to $\tau_m = 76$ msec. [Reprinted with permission from Mirau and Bovey (1986).]

B-DNA the pucker is 2′-endo. These features are more clearly observed in Figure 10.18 (Patel et al., 1982).

Observation of hydrogen-bonded imino-proton resonances in Watson–Crick base pairs (see Figure 10.14), which resonate at characteristically low fields (10–14 ppm) (Kearns et al., 1971; Patel and Tonelli, 1974), is diagnostic for the formation of double-helical DNA structures. ^{13}P NMR can distinguish between right-handed and left-handed DNA-duplex helices (Patel et al., 1982; Cohen and Chen, 1982). The effect on the ^{31}P NMR spectrum of poly(dG-dC)

Table 10.3 ■ Cross-Relaxation Rates, Calculated Interproton Distances, and X-Ray Distances for α-Helical PBLG[a]

Interaction	σ, sec^{-1}	r_{HH}, Å	
		From NOESY	Expected for α-helix (x-ray)
NH–αH	0.54	2.20	2.48
NH–βH	0.55	2.20	2.26
αH–βH	0.72	2.10	2.20

[a] Mirau and Bovey (1986).

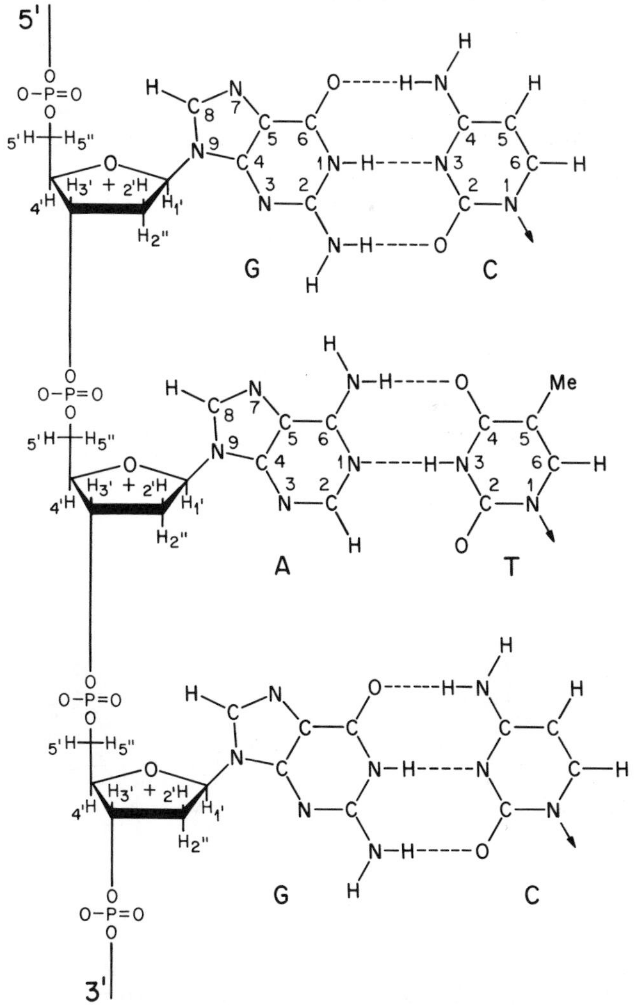

Figure 10.14 ■ Schematic diagram illustrating the chemical composition of DNA residues. Only one DNA strand is shown, through each of its bases is paired with its Watson–Crick complimentary base. [Adopted from Kearns (1987).]

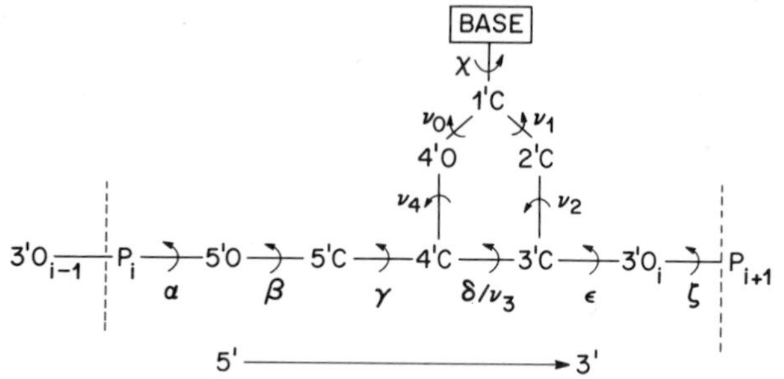

Figure 10.15 ■ Definition of torsion angles about the single bonds in a nucelotide residue *i* of a polynucleotide chain. [Adapted from Wüthrich (1986).]

during a salt-induced transition from the B- to the Z-DNA conformation is illustrated in Figure 10.19 (Patel et al., 1982). All the nucleotide residues in B-DNA adopt the 2′-endo-*anti* conformation, which leads to a single ^{31}P resonance, while in Z-DNA two ^{31}P resonances are observed, corresponding to the alternation of nucleotide residues between the 2′-endo-*anti* and 3′-endo-*syn* conformers.

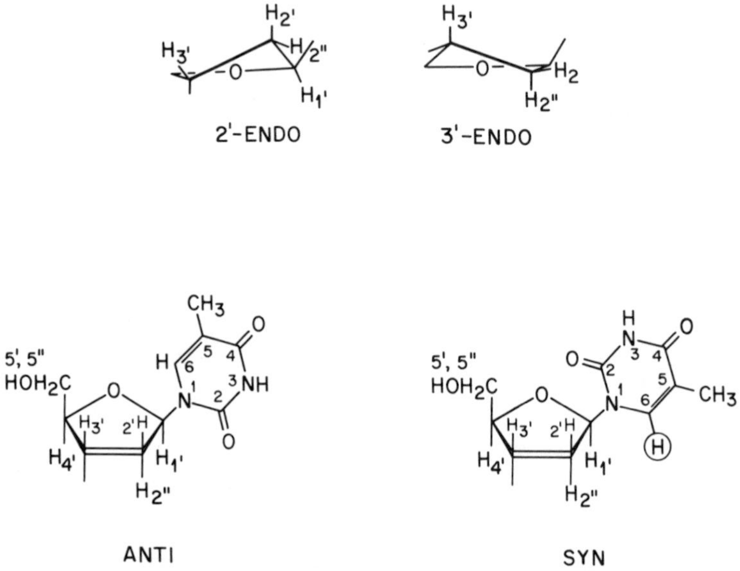

Figure 10.16 ■ Schematic drawings illustrating the two principal sugar puckers (top) and the two major torsion angles (*anti* and *syn*) about the glycosidic bond (bottom) connecting the sugars and bases in nucleotides. [Adapted from Kearns (1987).]

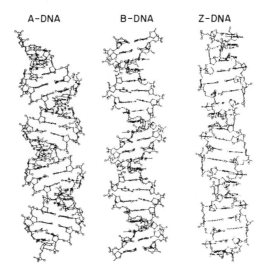

Figure 10.17 ■ Double helical structures of DNA. The right-handed helical A- and B-DNA and the left-handed Z-DNA are shown. [Reprinted with permission from Kearns (1987).]

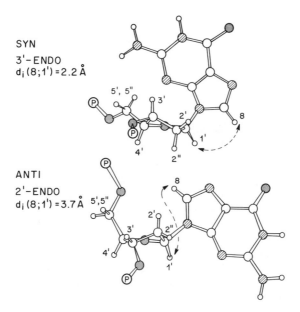

Figure 10.18 ■ Illustration of the dependence of the interproton distance between H-1' and H-8 on the glycosidic torsion angle χ. These two conformations are found for the G nucleotide residues of the Z from of $d(CGCGCG)_2$ (top) and in B-DNA (bottom). A-DNA is 3'-endo-*anti* (not shown). [Adapted from Patel et al. (1982).]

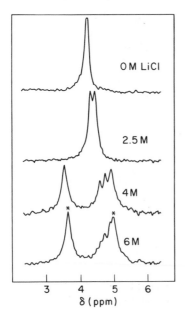

Figure 10.19 ■ 81-MHz proton-noise-decoupled ^{31}P NMR spectra of poly(dG–dC) at variable concentrations of LiCl in D_2O with 0.001 M EDTA, 0.01 M sodium cacodylate, and $pD = 7.2$. Asterisks indicate the ^{31}P resonance positions characteristic of Z-DNA, where δ is relative to trimethylphosphate. [Adapted from Patel et al. (1982).]

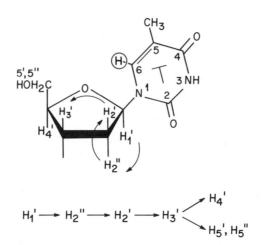

Figure 10.20 ■ Illustration of magnetization transfer within the sugar ring of one nucleotide. Magnetization introduced at H-1′ is first transferred to H-2″, followed by transfer to H-2′. Magnetization introduced at H-2′ will be transferred back to H-2″ and to H-3′ and from H-3′ to H-4′. [Adapted from Kearns (1987).]

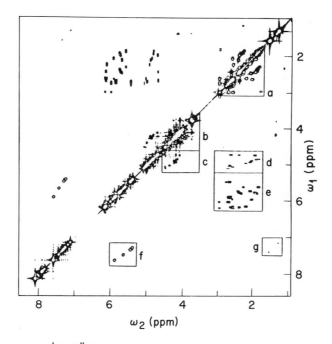

a. 2'H–2"H
b. 4'H–5'H, 4'H–5"H, AND 5'H–5"H
c. 3'H–4'H
d. 2'H–3'H AND 2"H–3'H
e. 1'H–2'H AND 1'H–2"H
f. 5H–6H OF C
g. FOUR–BOND CONNECTIVITY 5CH₃–6H OF T

Figure 10.21 ■ 500-MHz ¹H COSY spectrum of the DNA duplex d(CGCGAATTCGCG)₂. Regions of the spectrum labeled a–g correspond to the cross peaks generated by the spin–spin couplings listed below. [Adapted from Hare et al. (1983) and Wüthrich (1986).]

Figure 10.20 illustrates the transfer of magnetization between the spins within the sugar ring of one T nucleotide (Kearns, 1987). The 2D COSY spectrum of the DNA duplex d(CGCGAATTCGCG)₂ (Hare et al., 1983) is presented in Figure 10.21. The regions of the spectrum labeled a–g correspond to the cross peaks generated by the spin–spin couplings presented at the bottom of this figure. Regions f and g only occur in the pyrimidine nucleotides T and C and can be utilized for the identification of their resonances. We illustrate the sequential assignment of base and sugar proton resonances in DNA by means of 2D NOESY measurements with the schematic drawing of a single strand of DNA presented in Figure 10.22 (Kearns, 1987). For a right-hand helix with all bases in the *anti* conformation, a base proton of base

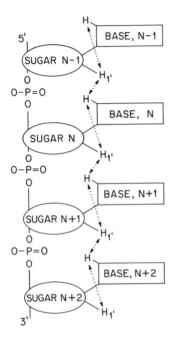

Figure 10.22 ▪ Diagram illustrating that the base proton (A or G H-8 or T or C H-6) in a regular right-handed DNA helix exhibits a NOE with the H-1′ sugar proton of the same nucleotide and with the H-1′ of the 5′ neighboring (N-1) nucleotide. [Reprinted with permission from Kearns (1987).]

N [H-6 (T or C) or H-8 (A or G)] can exhibit a NOE interaction with its own H-1′ sugar proton and the H-1′ sugar proton of the neighboring nucleotide N-1. The resonance from base $N + 1$ can be used to link the H-1′ resonances of nucleotides N and $N + 1$, and so forth down the DNA strand.

The pyrimidine H-6 and purine H-8 protons are close to the H-2′ proton on the same nucleotide in the *anti* conformation (see Figures 10.16 and 10.18), while in the *syn* conformation these base protons are near to H-1′ and far from H-2′. NOE effects can therefore be used to distinguish between these two glycosidic bond conformers. In the *anti* conformation these same base protons are closer to H-2′ than to H-3′ for the 2′-endo sugar pucker, but the opposite is true for the 3′-endo sugar pucker, and 2D NOE measurements are able to distinguish between them.

2D NOESY spectroscopy can also determine the sense of DNA helices. Internucleotide distances can be used to rigorously distinguish between left- and right-handed helices as illustrated in Figure 10.23 (Kearns, 1987). The nucleotides in both helices drawn there have the *anti* conformation, but no particular sugar pucker has been assumed. Note that in the right-handed stack of bases the H-6 or H-8 base proton may be close to the H-2″ (and possibly the H-1′ and H-2′) protons of the neighboring 5′ sugar. By contrast, the H-2′ and H-2″ protons are relatively remote from the base protons of either neighboring nucleotide in the left-handed helix, although the H-1′ proton of the neighboring 3′ sugar may approach the base protons.

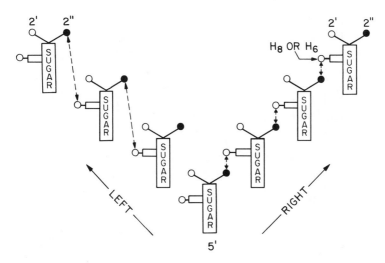

Figure 10.23 ■ Schematic drawing of the stacking of bases in right- and left-handed helices, where all bases are in an *anti* conformation, but no particular sugar pucker has been assumed. Note the differences in the base ↔ H-2″ intenucleotide interactions in the left- and right-handed helices. [Reprinted with permission from Kearns (1987).]

As an example, in the 2D NOESY spectrum of poly(dA–dT) · poly(dA–dT) presented in Figure 10.24 (Assa-Munt and Kearns, 1984), we observe that the cross peak between the TH-6 proton and the AH-1′ sugar proton is absent, while the interaction between the TH-6 and AH-2″ protons generates a cross peak which is easily visible in the same spectrum. This pattern of internucleotide proton interactions (see Figure 10.23) is diagnostic for a right-handed DNA helix. By comparing the relative intensities of the observed cross peaks with the interproton distances calculated for different DNA structures, Assa-Munt and Kearns (1984) concluded that poly(dA–dT) · poly(dA–dT) adopts the B-DNA structure in low-salt solutions.

It is hoped that this rather cursory description of polynucleotide microstructure as derived by means of 2D NMR techniques has at the very least demonstrated their utility. The reader is encouraged to refer to more complete discussions of the 2D NMR investigation of polynucleotide structure, such as those presented by Wüthrich (1986) and Kearns (1987).

10.4. Polysaccharides

The schematic representation of a polysaccharide in Figure 10.25, given as a succession of sugar moieties connected by glycosidic bonds, is deceptively simple. This deception is only partially uncovered by the detailed structures of

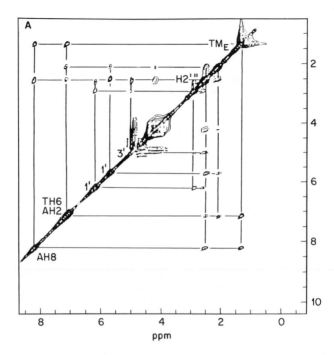

Figure 10.24 ■ 360-MHz pure absorption phase 2D NOE spectrum of poly(dA–dT) · poly(dA–dT) at 21°C obtained with a mixing time of 50 msec. [Reprinted with permission from Kearns (1987).]

the monosaccharides, or sugars, most commonly occurring in polysaccharides and presented in Figure 10.26 (MacGregor and Greenwood, 1980). Aside from the types and sequence of sugars constituting a given polysaccharide, there remain two additional structural characteristics originating from the manner in which two neighboring sugars are linked across the glycosidic bond.

The linking of two sugars across a glycosidic bond to form a disaccharide is illustrated in Figure 10.27. Because the anomeric C-1 carbon may be α or β, the glycosidic linkage may also be α or β. In addition, the anomeric C-1 carbon of one sugar may be linked to carbon atoms other than C-4, as depicted in Figure 10.27, to form $1 \rightarrow 2$, $1 \rightarrow 3$, and $1 \rightarrow 6$ glycosidic linkages as well. Because more than a single nonanomeric hydroxyl group may form a glycosidic bond to another sugar ring, branched structures may be formed in polysaccharides.

$$-\text{O}-\boxed{\text{SUGAR}}-\text{O}-\boxed{\text{SUGAR}}-\text{O}-\boxed{\text{SUGAR}}-\text{O}-$$

Figure 10.25 ■ Schematic drawing of a polysaccharide.

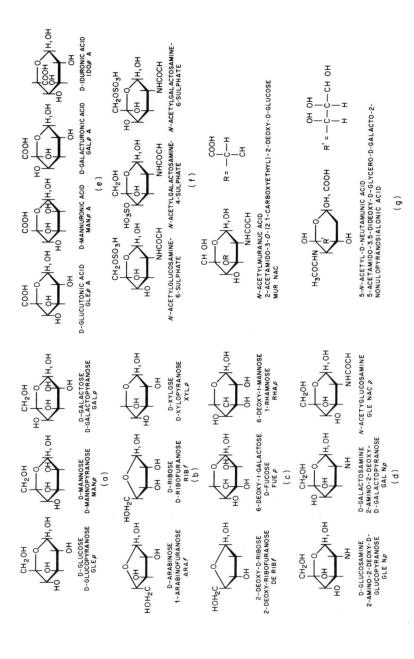

Figure 10.26 ■ Monosaccharides commonly occurring in polysaccharides: (a) aldohexoses; (b) aldopentoses; (c) deoxy sugars; (d) amino sugars and derivatives; (e) uronic acids; (f) sulfated sugars; (g) muramic acid and neuraminic acid derivatives. The designation H, OH at the C-1 anomeric ring carbon denotes either an α or a β sugar. [Reprinted with permission from MacGregor and Greenwood (1980).]

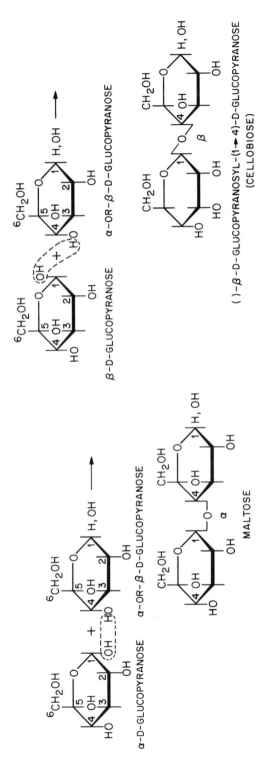

Figure 10.27 ■ Illustration of the formation of $1 \rightarrow 4$ α and β glycosidic bonds joining two sugars in disaccharides. [Adapted from MacGregor and Greenwood (1980).]

When the additional degrees of conformational freedom represented by the puckering of the six-membered sugar rings (pyranose sugars) and rotations about the C—O bonds of the glycosidic linkages are overlaid on top of the structural complexity of polysaccharides, it is not difficult to appreciate that to determine their three-dimensional structures is a formidable task. Because of this complexity, labeling techniques, comparison with model oligosaccharides, and observations of both ^{1}H and ^{13}C 1D and 2D NMR have been utilized to study the microstructures of polysaccharides (Bock and Thogerson, 1982; Breitmaier and Vöelter, 1987; Dabrowski, 1987).

Signal overlap in polysaccharide NMR spectra, especially ^{1}H NMR spectra, is a serious hindrance to their assignment. 2D NMR techniques are helpful in this regard, and early applications to polysaccharides appear encouraging (see Dabrowski, 1987). Considering the complexity of their NMR spectra and the

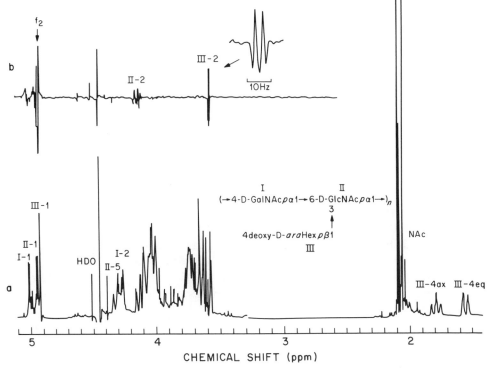

Figure 10.28 ■ 360-MHz ^{1}H NMR spectra of the polysaccharide produced by the bacterium Citrobacter PCM 1487: (a) resolution-enhanced and (b) spin-decoupling difference spectrum (irradiation indicated by the arrow marked f_2). [Reprinted with permission from Gamian et al. (1985).]

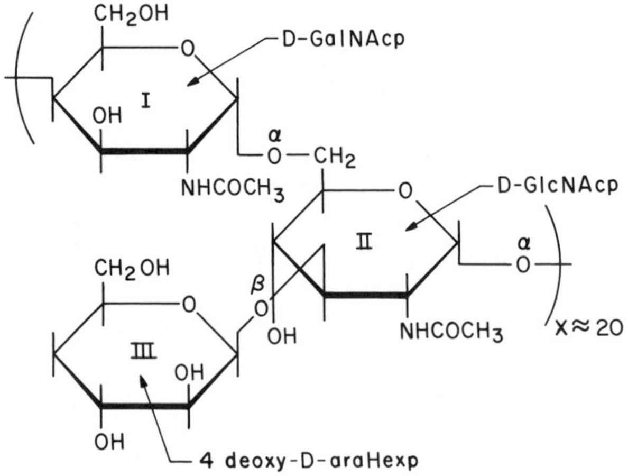

Figure 10.29 ■ Schematic drawing of the regular repeating structure of the polysaccharide produced by the bacterium Citrobacter PCM 1487. Measured molecular weights indicate an average of about 20 repeat units in each polysaccharide chain. [Reprinted with permission from Gamian et al. (1985).]

lack of an extensive literature regarding their investigation by 2D NMR, we will only briefly discuss a single illustrative example of the application of NMR to unravel some of the microstructural details of a polysaccharide.

Gamian et al. (1985) have applied 1D and 2D NMR methods to the polysaccharide produced by the bacterium Citrobacter PCM 1487. Three anomeric proton signals with intensity ratios $1:1:1$ are observed in the 1D spectrum presented in Figure 10.28. Based on their observed coupling constants $^3J_{1,2} = 3.6$ Hz, the anomeric configurations of the D-GalNAc(I) and D-GlcNAc(II) residues (see Figure 10.29) are both α, because a β-configuration would be expected to produce $^3J_{1,2} = 1.2$ Hz. These two sugar residues are readily distinguished by the large coupling constants of the H-4 proton in D-GlcNAc, which are absent in the D-GalNAc sugar residues.

The anomeric H-1 proton doublet at 4.905 ppm $(^3J_{1,2} = 1$ Hz) clearly belongs to the 4-deoxy-D-araHexp(III) residue, because it correlates (see the COSY spectrum in Figure 10.30), via H-2 and H-3, with the high-field signals of the two H-4 protons on the nonhydroxylated C-4 carbon of this residue. Based on the integrated intensity of the well-separated H-4 signals, each repeat unit in this polysaccharide contains only a single 4-deoxy-D-araHexp sugar residue as a sidechain.

For the sidechain sugar 4-deoxy-D-araHexp, which is also named 4-deoxy-D-altropyranose, either of the following two chair ring conformations is expected (Angyal, 1969; Paulsen and Freidman, 1972; DeBruyn et al., 1977):

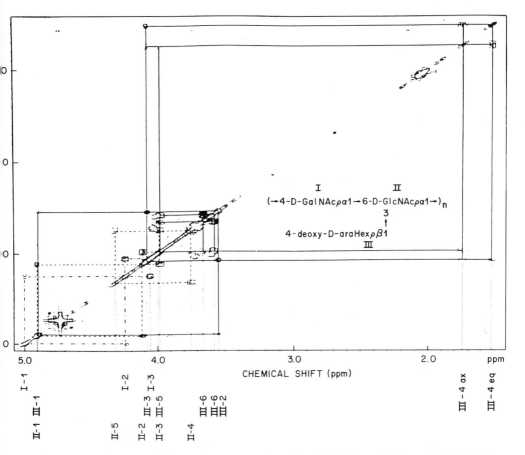

These two chair forms may be distinguished by their different $^3J_{2,3}$ coupling constants. This can be accomplished by a 2D J-resolved spectrum, or alternatively, by recording a single spin-decoupling difference spectrum as displayed in Figure 10.28(b). Irradiation of H-1 produces the H-2 difference signal displaying both the H-1 ⇔ H-2 and H-2 ⇔ H-3 coupling patterns. In this way

Figure 10.30 ■ 500-MHz 2D COSY ^{1}H NMR spectrum of the polysaccharide from Citrobacter PCM 1487. The spin-correlated connectivities are marked with dash-dot lines for GalNAc(I), (dashed) for GlcNAc(II), and (solid) for 4-deoxy-D-araHexp(III). [Reprinted with permission from Gamian et al. (1985).]

$^3J_{2,3} = 3$ Hz proves that the 4-deoxy-D-araHexp ring conformation is $^4C_1$. Finally, the β anomeric configuration was found for the 4-deoxy-D-araHexp residue, based on a comparison of its observed $^3J_{1,2} = 1$ Hz with those observed (Paulsen and Freidman, 1977) for the free α- and β-altropyranose sugars (3.6 and 1.2 Hz, respectively).

It is hoped that this brief account of the application (Gamian et al., 1985) of 1D and 2D NMR techniques to this relatively simple polysaccharide conveys some of their potential as powerful experimental probes of polysaccharide microstructure.

References

Altona, C. (1975). In *Structure and Conformation of Nucleic Acids and Protein–Nucleic Acid Interactions*, M. Sundaralingam and S. T. Rao, Eds., University Park Press, Baltimore.

Altona, C. and Sundaralingam, M. (1973). *J. Am. Chem. Soc.* **95**, 2333.

Altona, C., vanBoom, J., and Haasnoot, C. A. G. (1976). *Eur. J. Biochem.* **71**, 557.

Angyal, S. J. (1969). *Angew. Chem.* **81**, 172.

Assa-Munt, N. and Kearns, D. R. (1984). *Biochemistry* **23**, 791.

Bock, K. and Thogerson, H. (1982). In *Annual Reports on NMR Spectroscopy*, G. A. Webb, Ed., Vol. 13, Academic Press, New York, p. 1.

Bosch, C., Bundi, A., Oppliger, M. and Wüthrich, K. (1978). *Eur. J. Biochem.* **91**, 209.

Braun, W., Wider, G., Lee, K. H. and Wüthrich, K. (1983). *J. Mol. Biol.* **169**, 921.

Breitmaier, E. and Vöelter, W. (1987). *Carbon-13 NMR Spectroscopy*, Third Ed., VCH Publ., New York, Chapter 5.

Broersma, S. (1960). *J. Chem. Phys.* **32**, 1626.

Cohen, J. S. and Chen, C.-W. (1982). In *NMR Spectroscopy: New Methods and Applications*, ACS Symposium Series 191, Am. Chem. Soc., Washington, Chapter 13, p. 249.

Dabrowski, J. (1987). In *Two-Dimensional NMR Spectroscopy: Applications for Chemists and Biochemists*, W. R. Croasman and R. M. K. Carlson, Eds., VCH Publ., New York, Chapter 6, p. 349.

DeBruyn, A., Anteunis, M., and Beeumen, J. (1977). *Bull. Soc. Chim. Belg.* **86**, 259.

Dickerson, R. E. (1983). In *Nucleic Acids: The Vectors of Life*, B. Pullman and J. Jortner, Eds., Reidel, Dordrecht, p. 1.

Flory, P. J. (1969). *Statistical Mechanics of Chain Molecules*, Wiley-Interscience, New York, Chapter VII.

Gamian, A., Romanowska, E., Romanowska, A., Lugowski, C., Dabrowski, J., and Trauner, K. (1985). *Eur. J. Biochem.* **146**, 641.

Haasnoot, C. A. G., de Leeuw, F. A. A. M., de Leeuw, H. P. M., and Altona, C. (1981). *Org. Magn. Reson.* **15**, 43.

Hare, D. R., Werner, D. E., Chou, S. H., Drobny, G., and Reid, B. R. (1983). *J. Mol. Biol.* **171**, 319.

IUPAC-IUB Commission on Biochemical Nomenclature (1970). *J. Mol. Biol.* **52**, 1.

Kearns, D. R. (1987). In *Two-Dimensional NMR Spectroscopy: Applications for Chemists and Biochemists*, W. R. Croasmun and R. M. K. Carlson, Eds., VCH Publ., New York, Chapter 5.

Kearns, D. R., Patel, D. J., and Schulman, R. G. (1971). *Nature* **229**, 338.

MacGregor, E. A. and Greenwood, C. T. (1980). *Polymers in Nature*, Wiley, New York.

Mirau, P. A. and Bovey, F. A. (1986). *J. Am. Chem. Soc.* **108**, 5130.

Patel, D. J. and Tonelli, A. E. (1974). *Proc. Natl. Acad. Sci. U.S.A.* **71**, 1945.

Patel, D. J. Kozlowski, S. A., Nordheim, A., and Rich A. (1982). *Proc. Natl. Acad. Sci. U.S.A.* **79**, 1413.

Paulsen, H. and Freidman, M. (1972). *Chem. Ber.* **105**, 705.

Pohl, S. L., Birnbaumer, L., and Rodbell, M. (1969). *Science* **164**, 566.

Ramachandran, G. N., Ramakrishnan, C., and Sasisekharan, V. (1963). *J. Mol. Biol.* **7**, 95.

Sasaki, K., Dockerill, S., Adamiak, D. A., Tickle, I. J., and Blundell, T. (1975). *Nature* **257**, 751.

Tonelli, A. E. and Brewster, A. I. (1972). *J. Am. Chem. Soc.* **94**, 2851.

Tonelli, A. E. and Richard Brewster, A. I. (1973). *Biopolymers* **12**, 193.

Wider, G., Lee, K. H., and Wüthrich, K. (1982). *J. Mol. Biol.* **155**, 367.

Wüthrich, K. (1986). *NMR of Proteins and Nucleic Acids*, Wiley-Interscience, New York.

Wüthrich, K. Wider, G., Wagner, G., and Braun, W. (1982). *J. Mol. Biol.* **155**, 311.

Solid Polymers

11.1. Introduction

As first suggested and demonstrated by Schaefer and Stejskal (1976), combination of the techniques of high-power ^{1}H dipolar decoupling (DD), rapid magic-angle sample spinning (MAS), and cross-polarization (CP) of ^{1}H and ^{13}C nuclear spins permits the observation of high-resolution ^{13}C NMR spectra for solid samples (see Chapter 3). This is an important advancement in the characterization of polymers by NMR spectroscopy, because we may now observe their structures and dynamics in the solid state, where they are most often utilized.

Recording the CPMAS/DD ^{13}C NMR spectra of solid polymers makes possible the observation of the chemical shift (δ) and the relaxation parameters (T_1, $T_{1\rho}$, etc.) for each resolvable carbon nucleus. As we shall presently discuss, the observed chemical shifts provide information concerning the conformations and packing of solid polymers, while relaxation parameters, such as the spin–lattice relaxation time T_1, indicate their solid-state mobilities. Even though solid polymer chains are packed in close proximity and their ^{13}C chemical shifts can reflect these packing modes, we find, as observed in solution, that the intramolecular conformations adopted by solid polymers are the single most important influence on their observed ^{13}C chemical shifts.

We will illustrate the applicability of the conformationally sensitive γ-*gauche*-effect method (see Chapters 4 and 5) to the analysis of ^{13}C chemical shifts observed in the CPMAS/DD spectra of solid polymers. Such analyses often permit conclusions to be reached concerning the solid-state conformations of polymers and how they are affected by solid-state transitions (crystal–crystal, crystal–melt, crystal–liquid crystal). At the same time, measurement of the spin–lattice relaxation time T_1 for each observable resonance in a CPMAS/DD spectrum provides a measure of the dynamics of polymer motions in the solid state and can also be used to monitor solid-state polymer phase transitions.

In short, the combination of techniques signified by the acronym CPMAS/DD serve to conveniently obtain high-resolution NMR spectra of solid polymers. As a consequence, NMR spectroscopy, which is a sensitive probe of local molecular structure and dynamics, can now be applied to solid polymers, including those which are insoluble for reasons of cross-linking or high melting temperatures.

11.2. Solid-State Polymer Conformations

Most polymers with a regular repeating microstructure are able to crystallize both from solution and from the molten bulk. For polymers with irregular microstructures, such as atactic vinyl polymers (see Chapter 6), crystallization is usually not possible, and instead they coalesce into amorphous solids. Amorphous polymer chains are free to adopt any and all of the myriad of conformations they rapidly sample in solution. When a polymer crystallizes, however, usually only a single conformation is permitted for each chain in the crystalline domains of the sample.

Unlike low-molecular-weight molecules, polymers (macromolecules) generally do not crystallize completely. Degrees of crystallinity from 30 to 90% are commonly observed for crystalline synthetic polymers, with the remaining 10 to 70% of the sample in the disordered, amorphous state. Notable exceptions to this behavior are provided by the polydiacetylenes, which are obtained by the solid-state, topotactic polymerization of their monomer single crystals (Wegner, 1980), and the globular proteins, which fold upon themselves to form molecularly close-packed structures that are then able to organize in a three-dimensional crystalline lattice (Dickerson and Geiss, 1969). Thus most solid, synthetic polymers either are completely amorphous and conformationally diverse, or are two-phase materials with some chains, or portions of chains, fixed in a single crystalline conformation, while the remaining chains, or portions of chains, are amorphous.

What might we expect the high-resolution, solid-state CPMAS/DD ^{13}C NMR spectrum of a crystalline polymer to look like? Figure 11.1 presents a comparison of the CPMAS/DD ^{13}C NMR spectra of poly(butylene terephthalate) (PBT) recorded at ambient temperature and 105°C. [Peaks marked with sb are the spinning sidebands of the carbonyl and phenyl ring carbon resonances, and POM denotes the resonance of poly(oxymethylene) added to the rotor as a chemical-shift reference, i.e., 89.1 ppm from TMS (Earl and VanderHart, 1982).] Note the significant narrowing of resonances, especially those of the methylene carbons, in the high-temperature spectrum. This is likely a result of the increased mobility of the amorphous carbons at 105°C ($T_g = 55$°C), where they no longer efficiently cross-polarize. Thus, unlike the

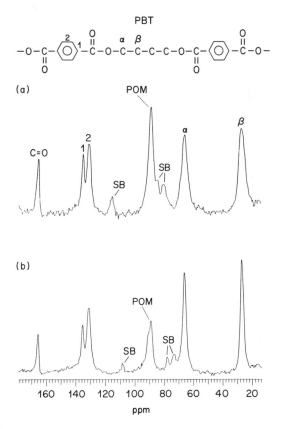

Figure 11.1 ■ CPMAS/DD spectra of α-PBT at ambient temperature (a) and 105°C (b). [Reprinted with permission from Gomez et al. (1988).]

spectrum recorded at ambient, only the crystalline carbons are observed at 105°C. At ambient temperature both the crystalline and amorphous carbons of PBT (50% crystalline) are sufficiently rigid to cross-polarize and contribute to the spectrum.

As seen in Figure 11.2, the amorphous carbons possess sufficient mobility at 105°C to be observed without cross-polarization. Comparison of the spectra recorded with (a) and without (b) cross-polarization reveals that the crystalline and amorphous carbon nuclei resonate at closely similar frequencies. For this reason the resonances observed at ambient temperature are broadened, because they possess significant contributions from both the rigid, amorphous carbons ($T_g = 55°C$) and the crystalline carbons.

Let us examine a crystalline polymer whose amorphous and crystalline resonances are more readily distinguished by the CPMAS/DD ^{13}C NMR techniques. Polyethylene crystallizes in the all-*trans*, planar zigzag conforma-

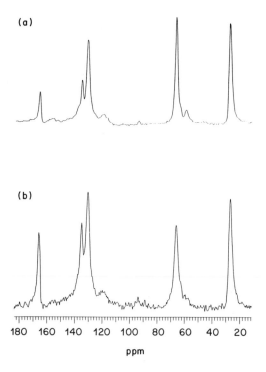

Figure 11.2 ■ CPMAS/DD (a) and MAS/DD (b) spectra of α-PBT measured at 105°C. [Reprinted with permission from Gomez et al., (1988).]

tion. In the melt, or in solution, where the conformational constraints of the crystalline lattice are removed, each polyethylene chain is free to adopt any of the conformations that are generated when each of its C–C bonds are either *trans* or *gauche*. Similar conformational flexibility is permitted for the amorphous portions of polyethylene chains in a solid crystalline sample.

In Figure 11.3(a) the CPMAS/DD ^{13}C NMR spectra of several polyethylene samples are presented (Earl and VanderHart, 1979). Despite the wide variations in their crystalline contents, which are a consequence of both their microstructures and the methods employed in their crystallization, each spectrum is characterized by an intense, narrow resonance accompanied by a broad, weak shoulder centered 2.5 ppm upfield. These resonances are assigned to the rigid crystalline and mobile amorphous portions of the samples, respectively. Application of a selective relaxation pulse sequence without cross-polarization enhances the observation of the mobile, amorphous component [see Figure 11.3(b); Axelson (1986)] because of the difference between the spin–lattice relaxation times of the crystalline and amorphous carbons.

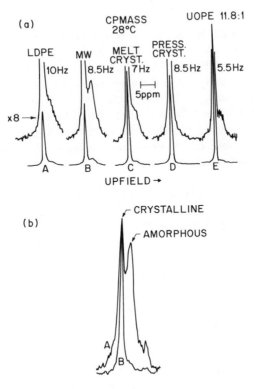

Figure 11.3 ▪ (a) CPMAS/DD ^{13}C NMR spectra of several polyethylene samples (Earl and VanderHart, 1979). (b) MAS/DD ^{13}C NMR spectra of polyethylene with CP (B) and without CP (A), but with a selective relaxation pulse sequence $(180°-\tau-90-T)_x$ (Axelson, 1986). [Adapted from Earl and VanderHart (1979) and Axelson (1986).]

Why do the conformationally disordered amorphous carbons in a semicrystalline sample of polyethylene resonate 2.5 ppm upfield from the crystalline carbons in the same sample? The major reason, aside from differences in the intermolecular packing of amorphous and crystalline polyethylene chains, is the different conformational environments experienced by the crystalline and amorphous carbon nuclei. In the crystalline portions of the sample, polyethylene chains are constrained by the lattice to adopt the all-*trans*, planar zigzag conformation. Amorphous chains, on the other hand, are free from the conformational constraints imposed by the crystalline lattice and are conformationally disordered, with appreciable numbers of their bonds in the *gauche* conformation.

In a disordered polyethylene melt or solution about 40% of the bonds are expected to adopt *gauche* conformations (Flory, 1969). If the amorphous chains in a semicrystalline sample of polyethylene are similarly disordered, then we might expect their carbon nuclei to be shielded by $(2)(0.4)(-5$ ppm) $= -4$ ppm compared to the all-*trans* crystalline carbons, which do not experience intramolecular γ-*gauche*-effect shielding. Clearly the small upfield peak observed in the CPMAS/DD ^{13}C NMR spectrum of polyethylene (see Figure 11.3) may be assigned to the amorphous portion of the sample, whose irregular conformations produce γ-*gauche* shielding of their carbon nuclei relative to the crystalline carbons fixed in the all-*trans* conformation.

Note that the observed difference between the ^{13}C chemical shifts for the amorphous and crystalline carbons in polyethylene (-2.5 ppm) is significantly smaller than that expected (-4 ppm) from the conformationally sensitive γ-*gauche* effect. Several reasons may be suggested for this difference. First, the comparison of the ^{13}C chemical shifts observed for the amorphous and crystalline phases in polyethylene with those predicted from their conformational differences by the γ-*gauche* effect ignores the possible effects of different modes of chain packing in the two coexisting phases. VanderHart (1981) has demonstrated that among the various crystalline phases of the *n*-alkanes, whose conformations are all *trans* like crystalline polyethylene, the chemical shifts of the central methylene carbons may differ by 1.3 ppm. It has also been suggested that the conformations of the amorphous chains in a semicrystalline polyethylene sample are not as diverse as those of molten polyethylene chains (see discussion concerning the morphology of polymer single crystals in Section 11.5.1).

Even though a quantitative discrepancy may exist between the observed and expected differences in the ^{13}C chemical shifts of crystalline and amorphous polyethylene, it will become further apparent that the local conformation of a solid polymer chain is the single most important influence on the chemical shifts observed for its solid-state ^{13}C resonances. Though the effects of interchain packing can at times measurably affect the ^{13}C chemical shifts observed for solid polymers, they generally play a secondary role compared to the local intrachain conformation.

The two stereoregular forms of polypropylene (PP), isotactic (*i*-PP) and syndiotactic (*s*-PP), both crystallize, but with distinct helical conformations (see Figure 11.4). A 3_1 helix with a regularly repeating ... *gtgtgt* ... conformation is assumed by crystalline *i*-PP, while *s*-PP adopts a ... *ggttggtt* ... 2_1 helical conformation in the crystal. The methylene carbons in crystalline *s*-PP are conformationally distinct. Half the methylene carbons, those lying along the helix interior, are in a *gauche* arrangement with both of their γ-substituents, while the exterior methylenes are *trans* to both of their γ-substituents. It is not surprising that in the high-resolution, solid-state ^{13}C NMR spectrum of *s*-PP (Bunn et al., 1981) two methylene carbon resonances appear and are separated by 8.7 ppm, or roughly two γ-*gauche* effects. Each of the methylene carbons in crystalline *i*-PP is *gauche* to one and *trans* to its other γ-substituent. We therefore expect only a single resonance for the methylene carbons in *i*-PP, as is observed (Fleming et al., 1980). Furthermore, the

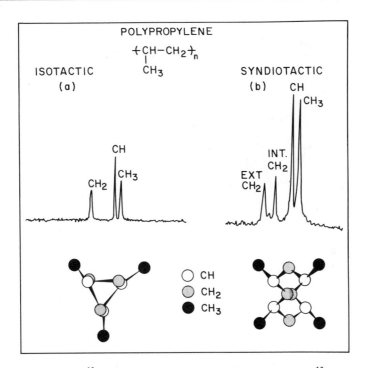

Figure 11.4 ■ Solid-state ^{13}C NMR spectra of polypropylene. (a) Solid state ^{13}C NMR spectrum of isotactic polypropylene [reproduced with modification from Fleming et al. (1980), by permission of the American Chemical Society] plus representation of the helical conformation of isotactic polypropylene. (b) Solid-state ^{13}C NMR spectrum of syndiotactic polypropylene, plus representation of the conformation of syndiotactic polypropylene. [Reproduced from Bunn et al. (1981) by permission of the Royal Chemical Society.]

methylene carbons in *i*-PP, which are shielded by a single γ-*gauche* effect, resonate midway between the two distinct methylene carbons in *s*-PP, which possess either two or no γ-*gauche* effects.

Normally isotactic poly(1-butene) crystallizes in a 3_1 ...$(gt)(gt)(gt)$... helical conformation, termed form I, much like *i*-PP, with *g* and *t* dihedral angles of 60° and 180°, respectively (Natta et al., 1960). However, above 90°C it prefers an 11_3 helix with *g* and *t* angles of 77° and 163°, respectively, called form II (Turner-Jones, 1963; Miyashita et al., 1974; Petraccone et al., 1976). When prepared in film form by solvent evaporation, poly(1-butene) crystallizes into a 4_1 helix with *g* and *t* angles of 83° and 159°, called form III (Zannetti et al., 1961; Danusso and Gianotti, 1963; Geacintov et al., 1963; Miller and Holland, 1964). Figure 11.5 presents the CPMAS/DD ^{13}C NMR spectra and Newman projections of forms I, II, and III of crystalline isotactic poly(1-butene). Notice that as the dihedral angles between CH and —CH$_2$— (between CH and —CH$_2$CH$_3$) open up from 60° (60°) to 77° (77°) to 83°

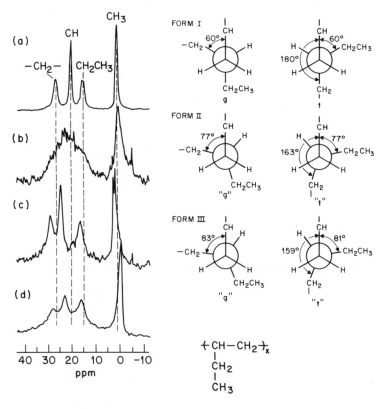

Figure 11.5 ■ Solid-state MAS/DD/CP 50.3-MHz carbon-13 spectra of poly(1-butene): (a) form I at 20°C; (b) form II at −60°C; (c) form III at −10°C; (d) amorphous at 43°C. The vertical dashed lines represent the peak positions of form I. The chemical-shift scale is referenced to the amorphous methyl resonance as 0.00 ppm. Newman projections of poly(1-butene) crystalline conformations are also presented. [Adapted from Belfiore et al. (1984).]

(81°) when isotactic poly(1-butene) is transformed from form I to form II to form III, the methine and methylene resonances move progressively downfield. Apparently the shielding γ-*gauche* effects experienced by the methine and methylene carbons in isotactic poly(1-butene) are reduced as the crystalline helices are opened up from form I (3_1) to form II (11_3) to form III (4_1). The methine carbon exhibits approximately twice the deshielding observed for the methylene carbons. This observation is consistent with the fact that both the backbone and sidechain methylenes shield the methine carbon, while only the methine carbon shields the methylene carbons (see Figure 11.5).

11.3. Interchain Packing in Solid Polymers

As mentioned in the preceding section, VanderHart (1981) observed the ^{13}C chemical shifts of the interior methylene carbons in crystalline *n*-alkanes to

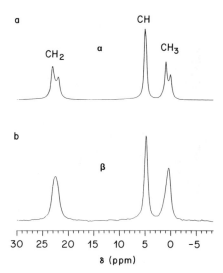

Figure 11.6 ■ CPMAS spectra of *i*-PP in (a) α-form and (b) β-form. [Adapted from Gomez et al. (1987a)].

depend on their mode of crystalline packing (unit cell), even though each *n*-alkane chain assumes the all-*trans* conformation in every crystalline polymorph. When *i*-PP is annealed above 150°C the stable α-form polymorph is obtained. On the other hand, unidirectional crystallization under a strong temperature gradient yields β-form *i*-PP. In both crystal forms the *i*-PP chains adopt the ...$(gt)(gt)(gt)$... 3_1 helical conformation. Their CPMAS/DD ^{13}C NMR spectra are presented in Figure 11.6, where in the α-form the methylene and methyl resonances are split into doublets separated by about 1 ppm. The interhelical packings of *i*-PP chains in the α- and β-form crystals are illustrated schematically in Figure 11.7. Note that helices of different handedness are packed in adjacent rows in the α-form, while clustering of helices of the same handedness occurs in the β-form crystals.

The ratio of intensities of the downfield to the upfield components is about 2 : 1 for both carbon types in the α-form spectrum. This corresponds to the ratio of nonequivalent packing sites, *A* and *B*, produced by pairing helices of opposite handedness [see Figure 11.7(a)]. The *A*-sites correspond to interhelical separations of 5.28 Å; the *B*-sites, to 6.14 Å. Methylene and methyl resonances in β-form *i*-PP nearly coincide with the corresponding *B*-site resonances in the α-form spectrum, probably because the interhelical separation between the clustered chains in the β-form crystals is similar (6.36 Å) to that between chains at the *B*-sites in the α-form crystals.

As a final example of the effects produced by interchain packing on the ^{13}C chemical shifts observed in high-resolution spectra of crystalline polymers, we may compare the CPMAS/DD ^{13}C NMR spectra of poly(trimethylene oxide) (PTO) and its cyclic tetramer *c*-(TO)$_4$ (Gomez et al., 1987b) shown in Figure 11.8. Both the polymer and the cyclic tetramer adopt the ...*ttgg*... conformation in their crystals. *c*-(TO)$_4$ actually adopts the (ttggtt$\bar{g}\bar{g}$)$_2$ conformation due

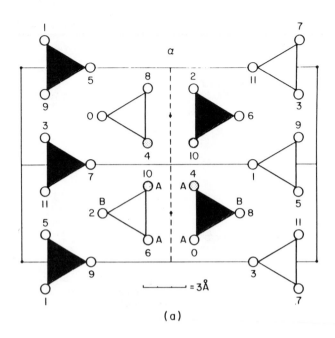

(a)

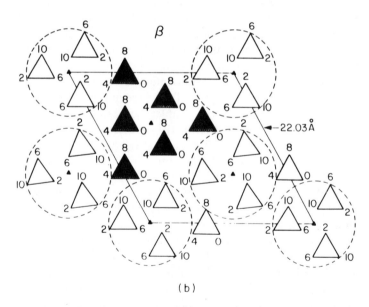

(b)

Figure 11.7 ■ Crystal structures of (a) α-form (Natta and Corradini, 1960b) and (b) β-form (Turner-Jones et al., 1964) i-PP. Full (RH) and open (LH) triangles indicate 3_1 helical i-PP chains of different handedness. A and B label the inequivalent sites discussed in the text, and are applicable to all three carbon types, because the CH–CH$_2$ bond is nearly parallel with the c-axis. Numerals at the triangle vertices indicate heights of methyl groups above a plane perpendicular to the c-axis in twelfths of c. The circles at the triangle vertices in (a) correspond to methyl carbons, and the cross-hatched and stippled pairs of circles correspond to the enmeshed A-site methyls. [Reprinted with permission from Gomez et al. (1987a).]

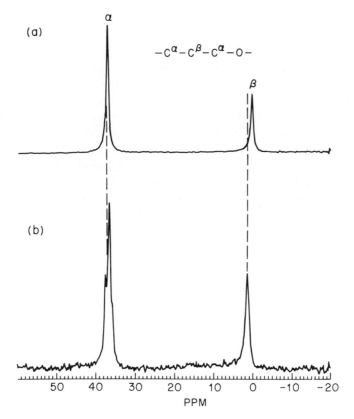

Figure 11.8 ▪ CPMAS spectra of PTO (a) and $c(TO)_4$ (b) recorded at room temperature, with no reference employed. The β-methylene carbon resonance in PTO is assigned 0 ppm. [Reprinted with permission from Gomez et al. (1987b).]

to its cyclic structure, but both α and β methylene carbons in crystalline PTO and $c\text{-}(TO)_4$ experience identical numbers and types of $\gamma\text{-}gauche$ interactions. Their ^{13}C chemical shifts differ by only 0.4 ppm ($\alpha\text{-}CH_2$'s) and 1.4 ppm ($\beta\text{-}CH_2$'s), even though the intermolecular packing of long PTO helices and compact $c\text{-}(TO)_4$ discs differ considerably (Tadokoro, 1979; Groth, 1971).

Based on these and other examples, we conclude that the effects of interchain packing on the ^{13}C chemical shifts observed in the high-resolution spectra of solid crystalline polymers, though significant and measurable, are not as large as the influence of the local polymer chain conformation operating via the $\gamma\text{-}gauche$ effect.

11.4. Molecular Motion in Solid Polymers

As noted previously (Sections 3.4 and 11.2), the measurement and appearance of a high-resolution, solid-state ^{13}C NMR polymer spectrum depend on the mobilities of its constituent carbon nuclei. In a semicrystalline polymer, for example, it is often possible, through judicious control of the measurement temperature, to separately observe the resonances of carbon nuclei belonging to either the crystalline or the amorphous portions of the sample. This separation is made possible by the different mobilities of the polymer chains in each phase. As a consequence, it is possible to learn something about the molecular motions occurring in solid polymers through observation by high-resolution, solid-state ^{13}C NMR spectroscopy and the dependence of the spectra on measurement parameters such as cross-polarization contact time, magic-angle spinning speed, ^{1}H dipolar decoupling period, etc. (Schaefer et al., 1975, 1977). A principal advantage of CPMAS/DD experiments is their high resolution, which allows relaxation data to be obtained for each resolvable carbon resonance in the polymer. Thus it may be possible to distinguish between motions occurring primarily in the polymer backbone and those in the side chains if they are not completely coupled.

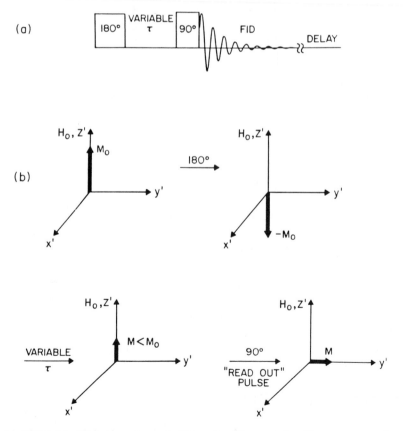

Figure 11.9 ■ (a) Pulse sequence and (b) vector diagrams for T_1-measurement by the inversion–recovery method. [Reprinted with permission from Bovey and Jelinski (1987).]

To measure the time T_1 it takes for the magnetization of a nuclear spin to return to its equilibrium value along the direction of the static magnetic field after being perturbed by the rf pulse, the inversion–recovery pulse sequence shown in Figure 11.9(a) is employed (Farrar and Becker, 1971). A 180° rf pulse tips the equilibrium net magnetization into the $-z$ direction, thereby inverting the magnetization. Relaxation via dipolar $^{13}C-^1H$ interactions, which are mediated by their motions, follows the 180° rf pulse and continues for a variable time τ. After time τ a 90° pulse is employed to detect the return of the magnetization along the $\mathbf{H}_0$-field direction $+z$ [see Figure 11.9(b)]. After waiting for the magnetization to return to its equilibrium value M_0 (a delay time of about $5T_1$), the pulse sequence is repeated. Data collected for a number of τ-values are collected, and $\log(M_0 - M)$ is plotted against τ. The spin–lattice relaxation time T_1 is obtained from the slope of the resultant straight line, i.e., $T_1 = -1/\text{slope}$.

Measurement of T_1's by the inversion–recovery method is practical only for mobile solids whose ^{13}C nuclei have short T_1's and whose resonances are observable without cross-polarization. However, Torchia (1978) has modified the usual cross-polarization pulse sequence (see Figure 3.8) to permit the measurement of ^{13}C spin–lattice relaxation times for rigid polymers whose spectra can only practically be obtained under the cross-polarization condition. This is achieved by inserting a variable delay τ between the spin locking

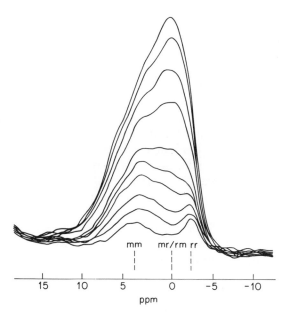

Figure 11.10 ■ Change of shape of α-methyl resonance as a function of τ in CP-T_1 experiment on a-PMMA. Spectra correspond to $\tau = 5$, 10, 20, 30, 60, 80, 100, 140, 200, and 300 msec, respectively, from the top to the bottom. [Reprinted with permission from Tanaka et al. (1988a).]

and decoupling of the proton spins and then repeating the pulse sequence, but this time using a $-90°$ proton rf pulse. ^{13}C T_1's may be obtained in this manner for rigid solids whose spectra are observable under the cross-polarization condition. By eliminating the second pulse sequence in the Torchia scheme (1978), spin–lattice relaxation times in the rotating frame, $T_{1\rho}$, may also be determined.

In Figure 11.10 the α-methyl portions of the CPMAS/DD ^{13}C NMR spectra of atactic poly(methyl methacrylate)

$$
a\text{-PMMA} = -(CH_2-\underset{\underset{CH_3}{|}}{\overset{\overset{\displaystyle CH_3}{\underset{\displaystyle |}{\overset{\displaystyle |}{O}}}}{C}})_x-
$$

recorded at room temperature in a T_1-experiment are presented. Note the dramatic change in shape of the α-CH_3 resonance produced by varying the delay time τ between the ^1H spin-locking and decoupling pulses in the CP–T_1 experiment (Torchia, 1978). It is apparent that this broad resonance may be decomposed into three peaks with relative chemical shifts of $+4.3$ ppm, 0 ppm, and -2.4 ppm, corresponding, respectively, to the triad stereosequences mm, mr (rm), and rr (Tanaka et al., 1988a).

The overall shape of the α-CH_3 resonance is sensitive to τ, because the spin–lattice relaxation time of α-CH_3 depends upon the stereosequence in which it resides. Assuming the intensity at $+4.3$ ppm is a combination of mm and mr (rm) resonances, the intensity at -2.4 ppm is a combination of rr and mr (rm) resonances, and only the mr (rm) resonance contributes to the intensity at 0 ppm, T_1's may be extracted for the α-CH_3 carbon in each of the mm, mr (rm), and rr triad stereosequences. The intensity-vs.-τ data at $+4.3$ ppm yield $T_1(mm)$ and $T_1[mr$ (rm)], at -2.4 ppm yield $T_1(rr)$ and $T_1[mr$ (rm)], and at 0 ppm yield $T_1[mr$ (rm)]. (See Figure 11.11.) Analysis of the intensity-vs.-τ data at all three resonance positions gives the same $T_1[mr$ (rm)].

At room temperature $T_1[mr$ (rm)] = 50 msec, $T_1(mm) = 400$ msec, and $T_1(rr) = 800$ msec for the α-CH_3 carbon in solid a-PPMA. Stereosequence-dependent spin–lattice relaxation has permitted a decomposition of the broad α-CH_3 resonance into three peaks corresponding to the mm, mr (rm), and rr triad stereosequences. The T_1's of the α-CH_3 carbons in solid a-PMMA depend strongly on tacticity and point to a remarkable sensitivity of α-CH_3-group rotation to stereoenvironment.

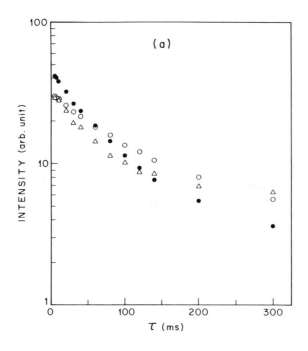

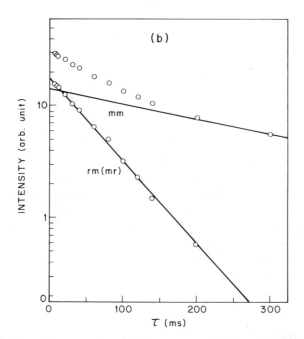

Figure 11.11 ■ (a) The change of intensity at three different chemical-shift positions as a function of τ. O: +4.3 ppm (*mm*); ●: 0 ppm (*rm, mr*); △: −2.4 ppm (*rr*). (b) Example decomposition of signal intensity at the relative chemical shift of +4.3 ppm (*mm*) into two processes. [Reprinted with permission from Tanaka et al. (1988a).]

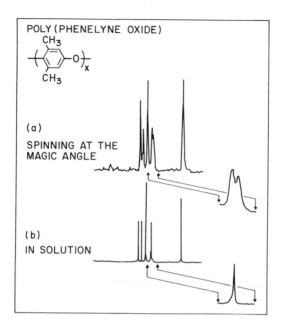

Figure 11.12 ■ Solid-state and solution-state ^{13}C NMR spectra of poly(phenylene oxide). (a) Solid-state spectrum obtained with CPMAS/DD. (b) Solution-state spectrum. [Figure reproduced with modification from Schaefer and Stejskal (1979), with permission of the copyright holder.]

If a pair or pairs of carbon nuclei that are magnetically equivalent in solution, because of rapid molecular reorientation, are induced by the solid state to be magnetically nonequivalent, then the motions of these carbons can be studied by following the temperature dependence of their solid-state CPMAS/DD spectra. Such an example is illustrated in Figure 11.12 (Schaefer and Stejskal, 1979). In the solid-state spectrum of poly(phenylene oxide) the protonated aromatic carbons appear as a doublet. This is a consequence of the fact that the C–O–C valence angle is about 120°, and in the absence of phenylene ring flips that are rapid on the NMR time scale the protonated aromatic carbons reside in environments different enough to give rise to separate isotropic chemical shifts. By raising the temperature and observing the coalescence of the doublet to a single resonance, as is observed in solution, it is possible to determine the activation energy for phenylene ring flips in the solid (Garroway et al., 1982).

Although not a high-resolution technique, observation of ^{13}C chemical-shift tensors in nonspinning spectra (see Chapter 3) can be utilized to elucidate motions in the frequency range of the MAS, i.e., kilohertz and slower. If in the molecular coordinate system the principal elements of the chemical-shift tensor σ can be assigned to specific directions, then the manner in which σ

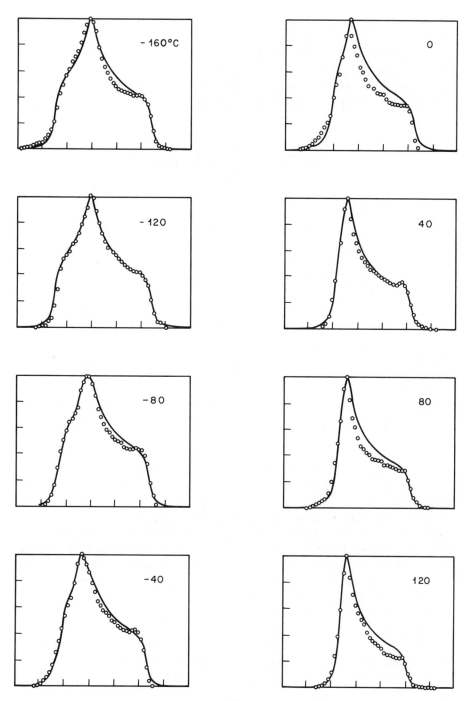

Figure 11.13 ▪ The lines are simulations of the ^{13}C CSA line shapes at several temperatures, and the points are taken from the spectra. [Reprinted with permission from O'Gara et al. (1985).]

changes with temperature, for example, may permit conclusions to be drawn regarding the molecular motion.

An example of such a motional analysis based on the temperature-dependent appearance of σ observed in nonspinning CP/DD spectra is summarized in Figure 11.13, where the chemical-shift tensor observed for the ^{13}C-enriched carbons (∗) in the polycarbonate of bisphenol-A

are presented at several temperatures (O'Gara et al., 1985). At low temperature σ exhibits the classic rigid tensor pattern (see Section 3.4.2), from which its principal components may be extracted. As the temperature is increased, 180°, or π, flips of the phenyl rings are permitted along with phenyl-ring rotations restricted to a small angular range. If the rate of π flips and the angular range of restricted rotation are both assumed to be temperature-dependent, then the chemical-shift tensors of the labeled ring carbons may be simulated from the principal components of σ as shown in Figure 11.13. In this way O'Gara et al. (1985) were able to draw conclusions about the rates and amplitudes of phenyl ring motion in the solid polycarbonate of bis-phenol-A.

Because the dipolar interactions between ^{1}H and ^{13}C nuclei (see Section 2.2) dominate the spin–lattice relaxation times (T_1 and $T_{1\rho}$) observed for carbon nuclei, it is difficult to extract models of molecular motions from these high-resolution data (Schaefer and Stejskal, 1979), especially for nonprotonated carbons, which only experience relaxation from dipolar interactions with nonbonded protons. Deuterium NMR, though not a high-resolution probe for solids, can be used to great advantage in the motional studies of solid polymers. The deuterium nucleus, $D = {}^2H$, has a spin of 1 and is quadrupolar as a consequence of its nonspherical charge distribution. The interaction of the quadrupole moment with the electric field gradient tensor of the C–D bond is dominant over the other NMR nuclear-spin interactions, such as scalar J-coupling, dipolar coupling, and chemical-shift anisotropy. It can be shown (Jelinski, 1986) that the frequencies of the two rf-induced transitions between the 0 and ±1 spin energy levels depend on the angle between the C–D bond and the applied external magnetic field.

Examples of deuterium line shapes expected for various rapid, anisotropic motions are presented in Figure 11.14. Comparison of these calculated line shapes with those observed in a selectively deuterated polymer solid often permits a detailed analysis of the types, frequencies, and amplitudes of the motions which are occurring there (Spiess, 1985; Jelinski, 1986).

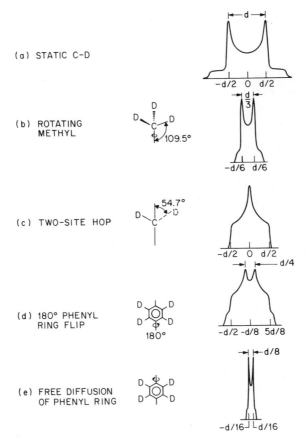

Figure 11.14 ■ Theoretical deuterium NMR lineshapes for various types of rapid ($\tau_c < 10^{-7}$ sec) anisotropic motions. [Reprinted with permission from Jelinski (1986).]

11.5. Application of CPMAS/DD ^{13}C NMR to Solid Polymers

11.5.1. Morphology and Motion in Polymer Crystals

When synthetic polymers are crystallized from their dilute solutions, small, thin, lozenge-shaped crystals, as shown schematically in Figure 11.15, are often formed. The thinness of the crystals is a consequence of polymer chain folding, which permits more than a single portion of most polymer chains to participate in the same crystal. The precise nature of the fold surface of polymer single crystals has been the focus of many investigations and much debate ever since Storks (1938), Keller (1957), Fisher (1957), and Till (1957) first demonstrated that the chains in polymer single crystals are folded at their surfaces.

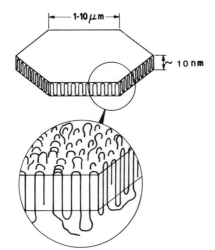

Figure 11.15 ■ Schematic drawings of a polymer single crystal grown from dilute solution. Note the thin (~ 100 Å), disclike habit of the crystal and its surface where the polymer chains fold to reenter the crystal.

One question has stimulated most of this work. Are the chain loops at the fold surface tight? If they are, then most of the polymer chains will reenter in an adjacent manner. If the folded polymer chain loops are loose, then a random reentry of polymer chains will result.

Schilling et al. (1983) addressed this problem with solution ^{13}C NMR. By treating single crystals of 1,4-*trans*-polybutadiene (TPBD) suspended in a liquid with *m*-chloroperbenzoic acid, which epoxidizes double bonds (see Figure 11.16), they were able to determine the average loop size on the crystal surface and the average crystal thickness or stem length of the chains in the

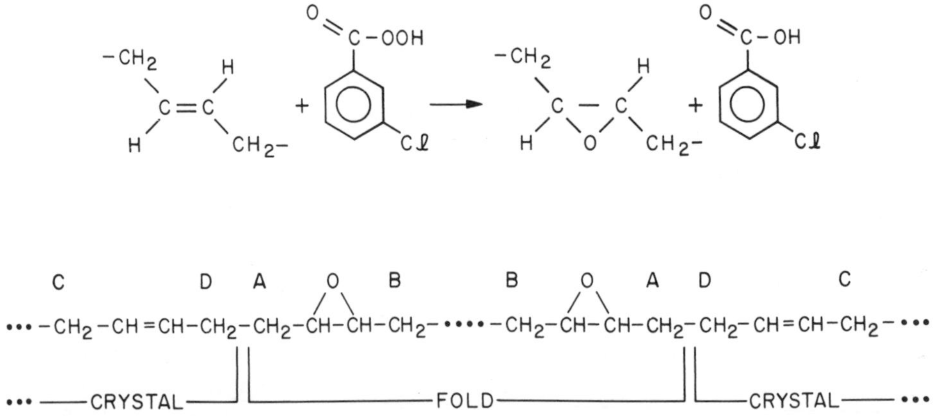

Figure 11.16 ■ Epoxidation of TPBD single crystals and the resulting block copolymers. [Reprinted with permission from Schilling et al. (1983).]

crystalline interior. This was accomplished by dissolving the single crystals with the epoxidized fold surfaces and analyzing the observed ^{13}C NMR spectra. If the epoxidation is confined to the fold surface, then the block copolymer depicted in Figure 11.16 is produced.

Determination of the intensities for the unique resonances of carbons A, B, C, and D results in an estimate of the fold and stem lengths of the unepoxidized TPBD single crystals. In this manner Schilling et al. (1983) found fold or loop lengths of 3–5 repeat units, which they interpreted as rather tight folds between predominantly adjacently reentering crystalline stems belonging to the same TPBD chains. These same TPBD single crystals were subsequently studied directly in the solid state by CPMASD/DD ^{13}C NMR (Schilling et al., 1984).

A quantitative MAS/DD ^{13}C NMR spectrum of TPBD single crystals is presented in Figure 11.17. The long time (200 sec) between signal accumulations assures observation of all the crystalline signal intensity. For both the aliphatic $-CH_2-$ and olefinic $-CH=$ carbons the upfield resonances

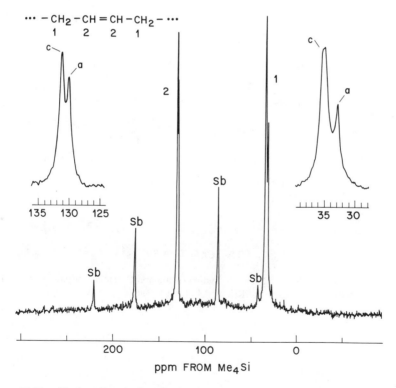

Figure 11.17 ■ Magic-angle-spinning, dipolar-decoupled, 50.3-MHz carbon-13 spectrum of 1,4-*trans*-polybutadiene, sample UH-29, using a 200-sec pulse interval without cross-polarization. [Reprinted with permission from Schilling et al. (1984).] c, a = crystalline, amorphous.

(a) CRYSTAL

$$E \quad s^{\pm} \quad t \quad s^{\mp} \quad E$$

$$\bullet\bullet\bullet -CH_2-CH_2-CH=CH-CH_2-CH_2-CH=CH-CH_2-CH_2-\bullet\bullet\bullet$$

(b) AMORPHOUS

$$E \quad c,s^{\pm} \quad t,g^{\pm} \quad c,s^{\pm} \quad E$$

$$\bullet\bullet\bullet -CH_2-CH_2-CH=CH-CH_2-CH_2-CH=CH-CH_2-CH_2-\bullet\bullet\bullet$$

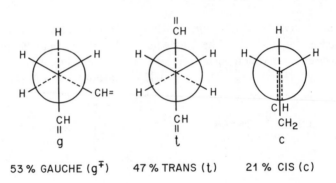

53 % GAUCHE ($g^{\mp}$)   47 % TRANS (t)   21 % CIS (c)

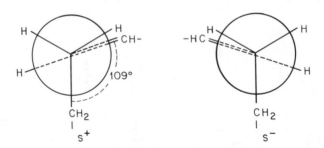

79 % EXACT SKEW ($s^{\pm}$)

Figure 11.18 ■ (a) Crystalline TPBD conformation (Natta and Corradini, 1960a). (b) Amorphous TPBD conformation according to the RIS model of Mark (1967).

correspond to the amorphous or fold surface material, as indicated in the MAS spectrum recorded (not shown) with low-power scalar decoupling, where only the mobile-amorphous-carbon resonances appear. The ratio of amorphous to crystalline peak intensities agrees well with the crystallinity determined by density measurements.

The TPBD chain conformations in the crystalline ($< 50°C$) and amorphous phases are indicated in Figure 11.18. Because all the bonds in the crystalline conformation are either *trans* or skew, neither the aliphatic nor the olefinic carbons are expected to experience intramolecular γ-*gauche* shielding effects. In the amorphous phase, however, the CH_2–CH_2 bonds are about 50% *gauche*, and the CH–CH_2 bonds are about 20% *cis*, leading to the expected shielding of both CH and CH_2 nuclei compared to their crystalline resonances. This expectation is indeed realized as indicated in Figure 11.17.

What is surprising, however, is the observation that the ^{13}C chemical shifts observed for the tightly folded carbons in the amorphous phase of single-crystal TPBD are virtually identical to those of a bulk, amorphous sample of TPBD (Jelinski et al., 1982), whose conformationally disordered chains are free of any crystalline constraints. Furthermore, the spin–lattice relaxation times T_1 measured for the bulk, amorphous sample and for the fold surface of the single crystal TPBD are also the same. Thus, both the chain motions and conformations in the fold surface of TPBD single crystals appear to be very similar to those in a bulk, amorphous sample despite the special constraints imposed on the rather tight (3–5 repeat units), adjacently reentering chain folds.

This apparent mystery can be at least partially resolved by application of variable-temperature CPMAS/DD ^{13}C NMR to TPBD single crystals. It has long been known (Natta et al., 1956; Natta and Corridini, 1960a) that TPBD exists in two crystalline polymorphs. At room temperature the chain conformation of form I is as indicated in Figure 11.18(a), but above about 75°C, though still packed in a hexagonal array, form-II TPBD crystals of lower density are formed (Suehiro and Takayanagi, 1970; Stellman et al., 1973; Evans and Woodward, 1978; De Rosa et al., 1986). In form II the TPBD chains are thought to be in a disordered state, as evidenced by the blurring of all nonequatorial reflections in its x-ray diffraction pattern (Suehiro and Takayanagi, 1970). A marked decrease in the second moment of the wide-line ^{1}H NMR spectrum of form-II TPBD (Iwayanagi and Miura, 1965) indicates the onset of molecular motion in the form-II crystals.

Suehiro and Takayanagi (1970) suggested that the form-II chains adopt a conformation similar to form I [see Figure 11.18(a)] except that the skew angle is decreased from ±109° (I) to ±80° (II) in order to reproduce the observed contraction in the chain-axis repeat distance seen by x-ray diffraction. They further proposed that the form-II chains undergo large torsional oscillations about the CH_2–CH_2 and CH–CH_2 bonds. Instead of these large torsional oscillations, Iwayanagi and Miura (1965) suggested that the form-II chains are rotating about their long axes. De Rosa et al. (1986) proposed a conformational disordering of TPBD chains in the form-II crystals. An equilibrating mixture of (a) and (b) conformations was suggested. This rapid conformational equilibrium produces a chain contraction along the fiber repeat which matches the observed reduction from form I to form II and leads to a 25% *cis* probability for the CH–CH$_2$ bonds:

$$(a) \quad E \quad s^{\pm}_{(90°)} \quad \dagger \quad s^{\pm}_{(90°)} \quad E$$
$$-CH = CH - CH_2 - CH_2 - CH = CH -$$
$$(b) \quad E \quad s^{\pm}_{(90°)} \quad \dagger \quad cis \quad E$$

CPMAS/DD ^{13}C NMR spectra, recorded with short contact times to depress observation of the amorphous fold surface carbons, are presented in Figure 11.19. Note the transition from form I to form II, where both carbon types in the form-II crystals resonate upfield from the corresponding form-I resonances. The CH and CH$_2$ carbons in form II resonate 1.2 and 1.8 ppm upfield, respectively, from form I, very close to those of the fold surface carbons, which are 1.2 and 2.4 ppm upfield from the form-I resonances. Individual form-II and fold surface methylene carbon resonances can be observed during the inversion–recovery T_1 measurements performed without cross-polarization. Figure 11.20 illustrates this separation, which is made possible by the different T_1's of form-II and fold surface TPBD.

Spin–lattice relaxation times T_1, measured for the crystalline carbons in form-I and -II TPBD under cross-polarization (Torchia, 1978), are presented in Table 11.1. Note the dramatic reduction in the T_1-values for both carbon types in the form-II crystals. The markedly greater shielding of the form-II stem carbons is difficult to understand in terms of the conformation proposed by Suehiro and Takayanagi (1970). Nor can torsional oscillations about the CH–CH$_2$ and CH$_2$–CH$_2$ bonds explain the large difference in carbon T_1-values for the form-I and -II crystalline stems. Both observations do seem consistent, however, with the disordered form-II conformations suggested by De Rosa et al. (1986), possibly combined with rotation about the long chain axes, as proposed by Iwayanagi and Miura (1965).

Additional insight into the nature of chain motion in form-II TPBD can be obtained from nonspinning ^{13}C NMR spectra, such as those presented in Figure 11.21. In Figure 11.21(a) the CP/DD spectrum of form I shows an axially asymmetric chemical-shift anisotropy [principal components σ_{11}, σ_{22}, σ_{33} were determined by Schilling et al. (1984)]. The dramatic narrowing of the powder pattern seen in the nonspinning spectrum of form-II TPBD recorded without cross-polarization [Figure 11.21(b)] indicates substantial motion, which, is, however, very anisotropic, because the chemical-shift tensor is not simply averaging to the isotropic value σ_i.

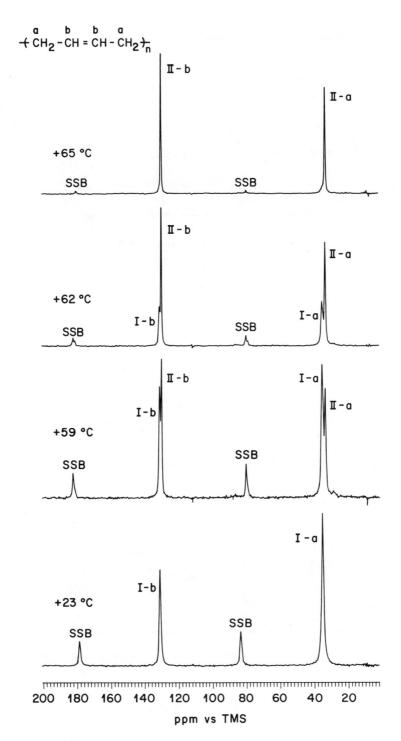

Figure 11.19 ■ ^{13}C NMR CPMAS/DD spectra, 50.31 MHz, of 1,4-*trans*-polybutadiene (I = form I; II = form II). [Reprinted with permission from Schilling et al. (1987).]

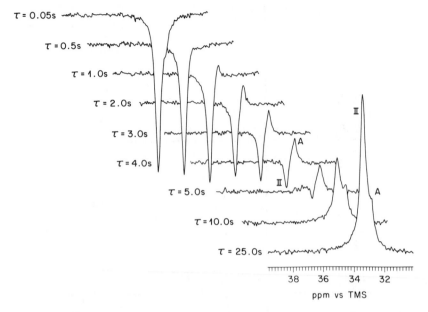

Figure 11.20 ■ ^{13}C NMR inversion–recovery spectra, 50.31 MHz, of the methylene carbons in 1,4-*trans*-polybutadiene at 70°C (II = form II; A = amorphous). [Reprinted with permission from Schilling et al. (1987).]

Table 11.1 ■ **Carbon-13 Cross-Polarization Spin–Lattice Relaxation Times for 1,4-*trans*-Polybutadiene**

Temp., °C	Form	T_1, sec			
		Stem		Fold	
		CH$_2$	CH=	CH$_2$	CH=
23.0	I	55, 130	53, 123	0.33^a	0.65^a
50.5	I	28, 69	40, 75		
60.0	I	23, 56	28, 66		
60.0	II	8.5	9.1		
70.0	II	10.5	12.2	~ 0.7^b	

^{a}Inversion–recovery measurement.
^{b}Estimate from inversion–recovery null point (Figure 11.20) (Schilling et al., 1987)

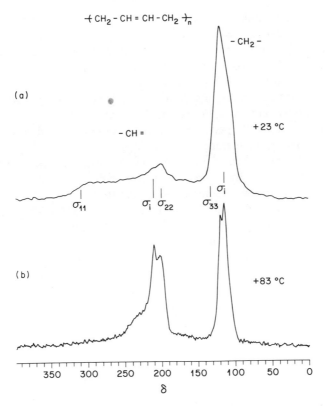

Figure 11.21 ■ ^{13}C NMR nonspinning DD spectra, 50.31 MHz, of 1,4-*trans*-polybutadiene: (a) form I with CP, and (b) form II without CP. [Reprinted with permission from Schilling et al. (1987).]

The solid–solid, form-I–form-II transition also occurs in the epoxidized TPBD single crystals (Schilling et al., 1987). Because epoxidation immobilizes the fold surface (Schilling et al., 1984), it is apparent that the fold surface is not involved in the form-I–form-II conversion. It is also likely that in form-II TPBD there is little motion along the direction of the crystalline stems, because such motion would require movement of the rigid, epoxidized fold surface into and out of the crystalline interior.

In light of the conformational and dynamic flexibility manifested by the TPBD chain stems in the form-II crystals, it does not seem as difficult to accept the observations that the relatively tightly folded surfaces of TPBD single crystals have conformational and motional characteristics similar to bulk, amorphous samples.

Ethylene–vinyl chloride (E–V) copolymers (see Chapter 9), which were made by the reductive dechlorination of poly(vinyl chloride) with tri-*n*-

butyltin hydride (Schilling et al., 1985), are semicrystalline when more than 60% of the chlorines are removed (Bowmer and Tonelli, 1985). For the crystalline E–V copolymers, X-ray diffraction shows significant expansion of the unit cell and intermolecular disorder, both of which increase with increasing V content until a change from the orthorhombic to a pseudohexagonal packing of chains occurs (Gomez et al., 1989). This implies that some of the V units, or Cl atoms, are incorporated into the crystal.

MAS/DD ^{13}C NMR spectra recorded above (86°C) and below (62° and 25°C) the melting temperature ($T_m = 78$°C) of the E–V-13.6 copolymer (13.6 mol% V) are presented in Figure 11.22. Spectra were recorded under conditions, established by T_1 measurements, where only the mobile amorphous carbons are observed (Gomez et al., 1989). At 86°C, where E–V-13.6 is molten, all carbons in the sample are observed in the MAS/DD spectrum. At 62°C, which is well below $T_m = 78$°C, E–V-13.6 is partially crystalline, and this is reflected in the spectrum recorded at this temperature. Differences in the intensity between the spectra recorded at 86 and 62°C reflect the amount of E–V-13.6 that has crystallized. More importantly, even the CHCl carbon resonance experiences an intensity loss upon crystallization, providing direct evidence that some Cl's are incorporated into the crystal. In fact, it was

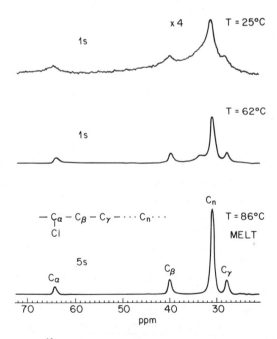

Figure 11.22 ▪ MAS/DD ^{13}C NMR spectra of E–V-13.6 recorded at 25, 62, and 86°C. [Reprinted with permission from Gomez et al. (1989).]

estimated that at least 20% of the Cl's in crystalline E–V copolymers find their way into the crystal interior (Gomez et al., 1989).

11.5.2. Solid–Solid Polymer Phase Transitions

In the course of discussing how solid-state ^{13}C NMR can be used to learn about the morphology and motions in polymer crystals, we briefly mentioned the form-I–form-II crystal–crystal transition observed in *trans*-1,4,-poly-butadiene. Here we expand on the utility of CPMAS/DD ^{13}C NMR in the study of solid-state polymer phase transitions.

The uniaxial extension of poly(butylene terephthalate) (PBT) fibers is accompanied by a reversible crystal–crystal transition (Boye and Overton, 1974; Jakeways et al., 1975, 1976; Yokouchi et al., 1976; Brereton et al., 1978). X-ray structural studies have been reported for both the relaxed α-phase and strained β-phase crystal forms (Yokouchi et al., 1976; Mencik, 1975; Hall and Pass, 1977; Desborough and Hall, 1977; Stambaugh et al., 1979; Hall, 1980). A *trans–trans–trans* conformation for the $-CH_2-CH_2-CH_2-CH_2-$ butylene glycol sequence is suggested by infrared and Raman spectroscopy for the strained β-phase PBT (Ward and Wilding, 1977), though the crystal structures proposed by Yokouchi et al. (1976) and Hall et al. (Hall, 1977; Hall and Pass, 1980) depart significantly from the extended, all-*trans* glycol structure. All crystal structures proposed for relaxed α-PBT approximate a *gauche–trans–gauche* glycol residue conformation.

The CPMAS/DD ^{13}C NMR spectra measured (Gomez et al., 1988) for α- and β-PBT at elevated temperature to remove contributions from the glassy, amorphous carbons (see Section 11.2) are presented in Figure 11.23. Note the nearly identical chemical shifts observed for the central methylene carbons in both α- and β-PBT. If the glycol portions of crystalline PBT were transformed from a *gauche–trans–gauche* (α) to the extended *trans–trans–trans* (β) conformation, then we would expect this to be reflected in the chemical shifts of the central methylene carbons due to γ-*gauche* effects.

In Figure 11.24 the central methylene ^{13}C chemical shifts and crystalline conformations determined for several PBT model compounds (Grenier-Loustalot and Bocelli, 1984) are compared with those observed for α- and β-PBT. Clearly the chemical shifts observed for the central methylene carbons in both α- and β-PBT are consistent with the extended, all-*trans* glycol conformation. Thus, it appears that during the strain-induced α-to-β crystal–crystal phase transition in PBT, the conformation of the butylene glycol portions of the PBT chains remain in the extended, all-*trans* conformation.

Poly(diacetylene)s (PDA) are unusual among synthetic organic polymers in that they can often be obtained as large single crystals by the solid-state, topochemical polymerization of single-crystal, substituted diacetylenes

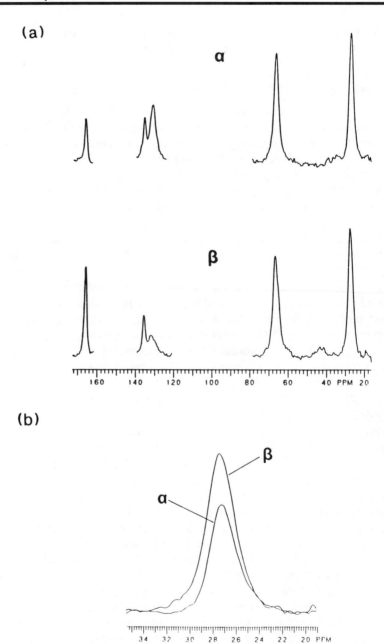

Figure 11.23 ■ CPMAS/DD spectra measured at 105°C for α- and β-PBT. Full spectra (a) and expansion of the central methylene-carbon regions (b). [Reprinted with permission from Gomez et al. (1988).]

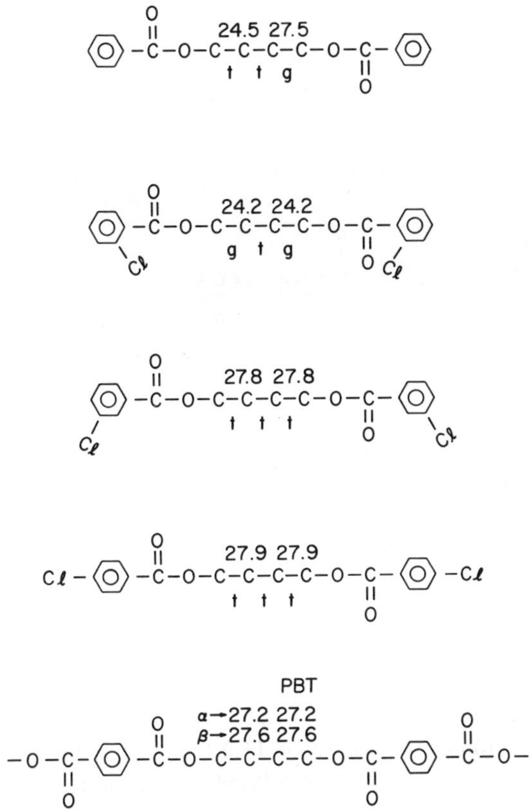

Figure 11.24 ■ Schematic drawings of the four PBT model compounds studied by Grenier-Loustalot and Bocelli (1984) using x-ray diffraction and high-resolution, solid-state ^{13}C NMR. The conformation of each glycol residue is indicated ($t = trans$, $g = gauche$), and the chemical shifts observed for the central methylene carbons are also listed. The structure of PBT is also presented, and the chemical shifts observed for the central methylene carbons in the α- and β-form crystals are indicated. [Reprinted with permission from Gomez et al. (1988).]

(Wegner, 1980). PDAs also exhibit interesting optical properties, such as large nonlinear optical responses and thermochromic phase transitions (Bloor and Chance, 1985; Chance, 1986). Their optical properties originate from their backbone structures, which consist of conjugated double and triple bonds as illustrated in Figure 11.25(a). Since the optical properties of PDAs directly reflect the electronic states of their backbones, it is very important to understand their backbone conformations.

The CPMAS/DD ^{13}C NMR spectra recorded at room and at elevated temperature for the PDA poly(ETCD) (see Figure 11.25) are presented in Figure 11.26. Above about 115°C poly(ETCD) transforms from the single-crystal blue to the single-crystal red phase. The ^{13}C chemical shifts observed in

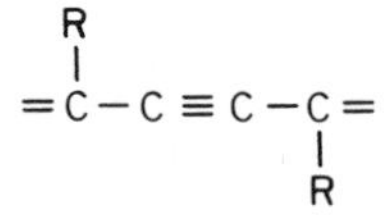

$$
\begin{array}{c}
R \\
| \\
=C-C\equiv C-C= \\
| \\
R
\end{array}
$$

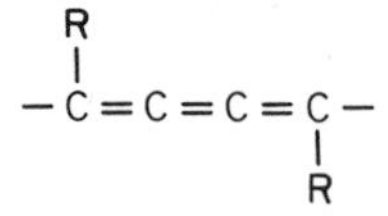

(a)

$$
\begin{array}{c}
R \\
| \\
-C=C=C=C- \\
| \\
R
\end{array}
$$

(b)

$$
R = CH_2 - CH_2 - CH_2 - CH_2 - OCONH - CH_2 - CH_3
$$
$$
\quad\;\; \alpha \qquad \beta \qquad \gamma \qquad \delta \qquad\qquad\qquad \epsilon
$$

(c)

Figure 11.25 ■ Acetylenic (a) and butatrienic (b) backbone structures, and the sidechain R (c) of poly(ETCD).

both phases are summarized in Table 11.2. There are no resonances corresponding to the butatrienic form of backbone conjugation [see Figure 11.25(b)], which would be expected to occur between 136 and 171 ppm (van Dongen et al., 1973; Sandman et al., 1986) in either phase of poly(ETCD). Judging from the ^{13}C chemical shift of the C=O carbon (Saito, 1986), we also conclude that the inter-sidechain hydrogen bonds between urethane groups are maintained through the thermochromic blue-to-red phase transition (see Figure 11.27).

The only significant changes in the CPMAS/DD ^{13}C NMR spectra of poly(ETCD) caused by its thermochromic phase transition are the 4-ppm upfield shift observed for the backbone $-C\equiv$ resonance and the 2-ppm downfield shift observed for the β, γ-CH$_2$ sidechain carbons (see Table 11.2 and Figures 11.25 and 11.26). Further solid-state ^{13}C NMR studies of several additional PDAs with a variety of sidechains, some with and others without the capability of forming an inter-sidechain hydrogen-bonded network (Tanaka et al., 1989b), revealed that all the blue-phase PDAs have $-C\equiv$ resonances at about 107 ppm, while the red-phase PDAs all show $-C\equiv$ resonances at ca. 103 ppm. This was interpreted as resulting from a planar-to-slightly-nonpolar conformational transition of the PDA backbones. Small rotations of opposite sign about the $\equiv C-C=$ bonds in the acetylenic PDA structure [see Figure 11.25(a)] are thought to accompany the thermochromic phase

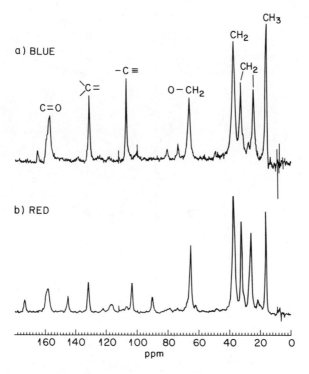

Figure 11.26 ■ CP/MAS/DD ^{13}C NMR spectra of poly(ETCD) in the blue and red phases. [Adapted from Tanaka et al. (1989a).]

Table 11.2 ■ ^{13}C Chemical Shifts of Poly(ETCD) vs. TMS

	Chem. shift, ppm	
Carbon	Blue Phase (Low T)	Red Phase (High T)
C=O	159.3	—
	157.5	158.3
> C=	131.6	132.0
—C≡	107.4	103.6
δ-CH$_2$	66.6	65.5
α-CH$_2$	37.3	37.8
ϵ-CH$_2$	32.9	32.6
β, γ-CH$_2$	24.5	26.4
CH$_3$	16.2	16.7

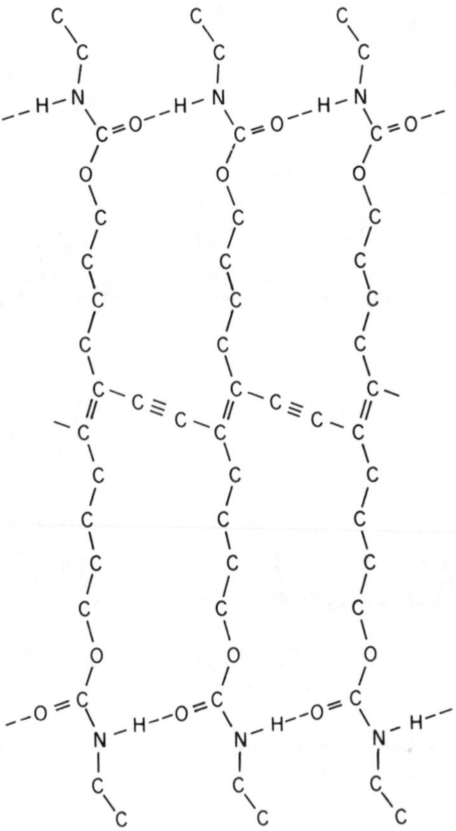

Figure 11.27 ■ Schematic diagram for the intramolecular structure of poly(ETCD) with hydrogen bonding. Protons are not drawn in this figure, and a double dash represents the hydrogen bond. This diagram is an approximation of the crystal structure reported (Enkelmann and Lando, 1978) for poly(TCDU), which differs in molecular structure from poly(ETCD) only by the substitution of a phenyl ring for the ethyl group in the sidechains. [Reprinted with permission from Tanaka et al. (1987).]

transitions observed in many PDAs, and to be responsible for the dramatic reduction in the length of backbone π-electron conjugation which occurs there.

Unlike the poly(ETCD) backbone, the sidechains appear to be more extended in the high-temperature red phase. ^{13}C chemical shifts observed for the central methylene carbons (β, γ) of the poly(ETCD) sidechains were compared with the ^{13}C chemical shifts observed for structurally similar methylene carbons in other crystalline polymers and model compounds whose solid-state conformations are known from x-ray diffraction studies. This comparison is illustrated in Figure 11.28, where it appears that the

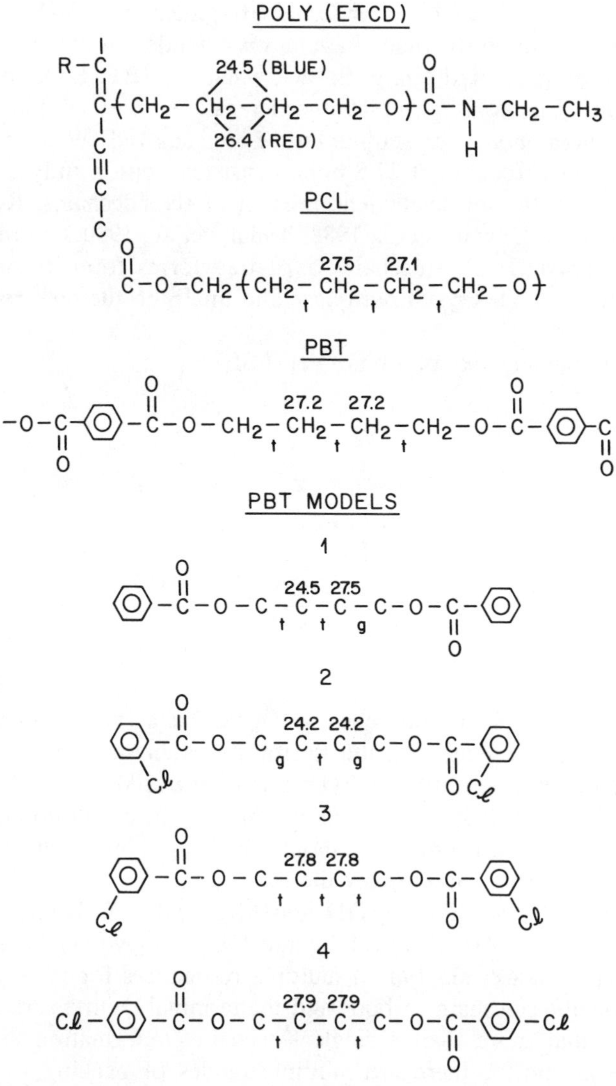

Figure 11.28 ■ ^{13}C chemical shifts for the β and γ CH$_2$ carbons observed in the solid state for poly(ETCD), PCL, PBT, and PBT-model compounds. Conformations ($t = trans$, $g = gauche$), as determined by x-ray diffraction (Tadokoro, 1979; Grenier-Loustalot and Bocelli, 1984) are indicated below each of the central C–C bonds. [Reprinted with permission from Tanaka et al. (1989a).]

—CH$_2$—CH$_2$—CH$_2$—CH$_2$— sidechain fragment in poly(ETCD) adopts the $gt\bar{g}$ conformation in the blue phase and is extended to the nearly all-*trans*, *ttt* conformation in the red phase. In fact, when poly(ETCD) is recrystallized from its melt (Tanaka et al., 1989c), where all memory of the constraints imposed by the monomer crystalline lattice are removed (Wegner, 1980), the β, γ-CH$_2$ carbons resonate at 27.5 ppm, consistent with a fully extended, *ttt* conformation for the tetramethylene portion of its sidechains. Recent x-ray diffraction studies (Downey et al., 1988; Tanaka et al., 1989c) corroborate this extension of poly(ETCD) sidechains as it transforms from its single-crystal blue phase to its single-crystal red phase and finally to the melt-recrystallized material.

Poly[bis(4-ethylphenoxy)phosphazene] (PBEPP),

PBEPP

is typical of many polyphosphazenes which exhibit a crystal-to-liquid-crystal transition well before achieving the isotropic molten state (Sun and Magill, 1987). For PBEPP this transition, T(1), occurs near 100°C. By observing the CPMAS/DD ^{13}C NMR spectra of PBEPP through the crystal-to-liquid-crystal transition, we may learn something about the changes in sidechain conformation and mobility which accompany the transition.

In Figure 11.29 the CPMAS/DD and MAS/DD ^{13}C NMR spectra observed for PBEPP at 24°C [below T(1)] and 120°C [above T(1)] are displayed. We observe (Tanaka et al., 1988b) multiple resonances for three of the four different aromatic sidechain carbons and the terminal methyl carbon at 24°C. This implies that more than a single sidechain conformation exists in the PBEPP crystal and/or there are multiple modes of packing its crystalline sidechains.

In Table 11.3 we present the CP spin–lattice relaxation times T_1 measured for crystalline and liquid-crystalline PBEPP. Aside from the reduction in T_1's expected as PBEPP transforms from the rigid-crystalline to the mobile liquid-crystalline phase, note that two of the sidechain aromatic carbons have relatively small T_1's even in the crystalline phase. In fact, $T_1(C_b) = T_1(C_c) = T_1(CH_3)$. This behavior implies that the phenyl rings are rotating or flipping about their 1,4-axes even in the crystalline phase. However, because we observe multiple resonances for the crystalline aromatic carbons, this motion is likely slow on the NMR time scale, i.e., $> 10^{-3}$ sec.

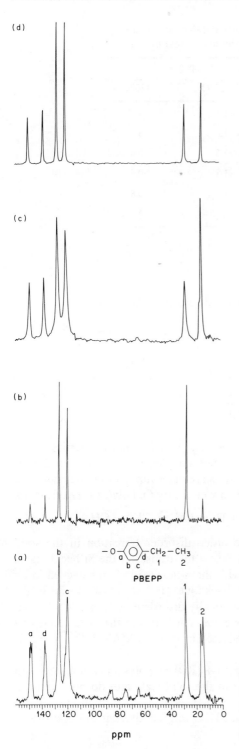

(d)

(c)

(b)

(a)

$-O-\underset{\underset{b\ c}{a}}{\bigcirc}-\underset{d}{CH_2}-\underset{2}{CH_3}$

PBEPP

Figure 11.29 ■ ^{13}C spectra below and above T(1). (a) CP MAS/DD spectrum at 24°C. (b) CP MAS/DD spectrum at 120°C. (c) MAS/DD spectrum at 24°C. (d) MAS/DD spectrum at 120°. [Reprinted with permission from Tanaka et al. (1988b).]

Table 11.3 ■ Spin–Lattice Relaxation Times T_1
for polybis(Ethyl phenoxy)phosphazene

	T_1, sec	
Carbon	$T = 25°C$	$100°C$
C_a	17	4
C_d	15	3
C_b	1.5	0.6
C_c	1.5	0.5
CH_2	10	0.8
CH_3	2	2

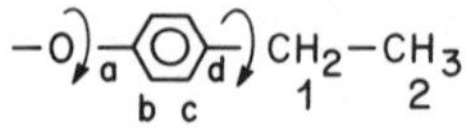

$$-O\!\!\left(\!\!\underset{a}{}\!\!\underset{b\ c}{\bigcirc}\!\!\underset{d}{}\!\!\right)\!\!\underset{1}{CH_2}-\underset{2}{CH_3}$$

11.6. Other Nuclei Observed in Solid-State Polymer Spectra

11.6.1. CPMAS/DD ^{29}Si NMR

The polysilanes

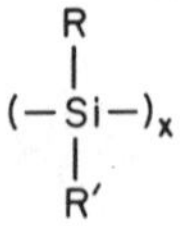

$$(-\underset{\underset{R'}{|}}{\overset{\overset{R}{|}}{Si}}-)_x$$

are an interesting class of inorganic polymers with unusual properties. Their electronic absorption properties are strongly coupled to the conformation of the Si backbone, because the Si σ-electrons are delocalized along the backbone (Harrah and Ziegler, 1985, 1987; Kuzmany et al., 1986). Poly(di-n-hexylsilane),

$$R = R' = CH_2 - CH_2 - CH_2 - CH_2 - CH_2 - CH_3$$

(PDHS), undergoes a thermochromic order–disorder transition in the solid state at 42°C (Schilling et al., 1989a). Below the transition the Si backbone is in the all-*trans* planar conformation with the n-hexyl sidechains packed in an ordered array perpendicular to the backbone (I). Above 42°C both the backbone and sidechains are conformationally disordered (II). The conformational differences between the backbone and sidechains in the ordered (I) and disordered (II) forms of PDHS are reflected in their CPMAS/DD ^{29}Si NMR spectra as evidenced in Figure 11.30.

The di-n-butyl (PDBS) and di-n-pentyl (PDPS) polysilanes adopt a 7/3 helical conformation (30° from *trans*) in the solid state and do not exhibit a solid-state thermochromic transition (Miller et al., 1987; Schilling et al., 1989b). However, it is possible to form PDBS with an all-*trans* Si backbone in

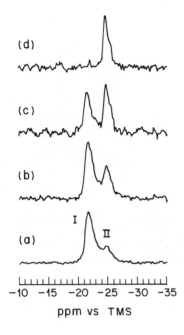

(d)

(c)

(b)

I

II

(a)

-10 -15 -20 -25 -30 -35

ppm vs TMS

Figure 11.30 ■ 39.75-MHz ^{29}Si CPMAS/DD spectra of solid-state crystallized PDHS at (a) 25, (b) 39.5, (c) 41.5, and (d) 44.3°C. I = resonance of phase I; II = resonance of phase II. [Reprinted with permission from Schilling et al. (1986).]

the solid state either by precipitation from dilute solution at very low temperatures (-78°C) or by application of modest pressures (2.6 kbar) to thin film samples (Walsh et al., 1989). Figure 11.31 compares the CPMAS/DD ^{29}Si NMR spectra of PDBS precipitated at low temperature with the spectra of PDBS and PDHS both precipitated at room temperature. It appears that the broad resonance of PDBS precipitated at low temperature consists of three components: (i) the all-*trans* form, (ii) the disordered form, and (iii) the 7/3 helical form (Walsh et al., 1989).

11.6.2. MAS/DD ^{31}P NMR

Because the ^{31}P nucleus is abundant, it is not necessary to accumulate many scans to achieve a sufficient signal-to-noise ratio. Thus, even though the spin–lattice relaxation times of ^{31}P nuclei in solids are long, cross-polarization of magnetization between ^{1}H and ^{31}P nuclei is not required to obtain a solid-state spectrum in a reasonable period of time. Figure 11.32 presents the MAS/DD ^{31}P spectra of the polyphosphazene PBEPP (see Section 11.5.2) as it is heated and cooled through its T(1) transition between crystal and liquid-crystal phases. Below T(1) (< 100°C), three components are evident in the spectra and were assigned to the crystalline, interfacial, and amorphous components of the sample from low to high field. Above T(1), only a single liquid-crystalline component appears.

In the nonspinning ^{31}P spectra presented in Figure 11.33, though considerable narrowing is observed above T(1), chemical shift anisotropy is still

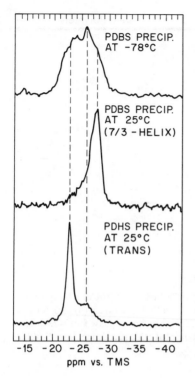

Figure 11.31 ■ ^{29}Si solid-state NMR spectra (recorded with magic-angle spinning, cross-polarization, and dipolar decoupling) of PDBS and PDHS at −40°C. The broad resonance of PDBS precipitated at low temperature arises from the superposition of resonances for three general backbone structures: all-*trans* (downfield shoulder), helix (upfield shoulder), and disordered (center peak). [Reprinted with permission from Walsh et al. (1989).]

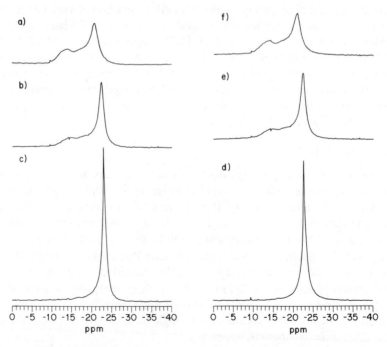

Figure 11.32 ■ ^{31}P NMR MAS/DD spectra of poly[bis-(4-ethylphenoxy)phosphazene] at (a) 23, (b) 80, and (c) 120°C in the heating process and at (d) 100, (e) 60, and (f) 23°C in the cooling process. [Reprinted with permission from Tanaka et al. (1989d).]

242

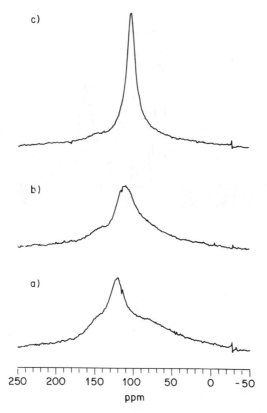

Figure 11.33 ■ Nonspinning ^{31}P NMR spectra of poly[bis-(4-ethylphenoxy)phosphazene] at (a) 23, (b) 70 (cooling), and (c) 90°C (cooling). [Reprinted with permission from Tanaka et al. (1989d).]

evident in the liquid-crystalline state. The line width is much broader than with MAS (15 ppm vs. < 1 ppm), suggesting that the motion is neither isotropic nor as rapid as in a true liquid.

11.6.3. CPMAS/DD ^{15}N NMR

Like ^{13}C, ^{15}N is a spin-$\frac{1}{2}$ nucleus, though both its natural abundance and its magnetic moment are less than for ^{13}C and make its observation more difficult. Consequently, ^{15}N-enriched samples are most often employed. However, it has recently been demonstrated (Shoji et al., 1987; Weber and Murphy, 1987; Mathias et al., 1988; Powell et al., 1988) that CPMAS/DD ^{15}N NMR spectroscopy of solid polymers is feasible at natural abundance. For example, the ^{15}N resonances of the alternating copolyamide poly(p-benzamide-*alt*-caproamide) are well separated (20 ppm) (see Figure 11.34).

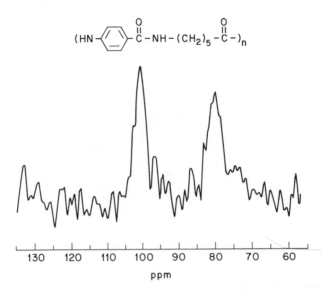

Figure 11.34 ■ ^{15}N CPMAS spectrum of poly(p-benzamide-*alt*-caproamide) alternating copolymer. [Reprinted with permission from Powell et al. (1988).] δ^{15}N relative to crystalline glycine.

Nylon-6,

$$(-NH-(CH_2-)_5-\overset{\overset{\displaystyle O}{\|}}{C}-)$$

is known (Holmes et al., 1955; Arimoto, 1964) to crystallize into two polymorphs, the α and γ forms. The solid-state ^{15}N NMR spectra of nylon-6 shown in Figure 11.35 exhibit a clear dependence on the form in which it crystallizes. Over 5 ppm separate the ^{15}N resonances of the α and γ crystalline forms of nylon-6. Judging from the magnitude of the observed ^{15}N chemical-shift separation, it seems likely that conformational differences between the nylon-6 chains in the α and γ phases are at least partially responsible for the observed ^{15}N chemical shifts.

11.7. Concluding Remarks

Before closing this final chapter, it should be mentioned that high-resolution solid-state NMR studies similar to those described here for synthetic polymers have also been carried out for biological macromolecules. CPMAS/DD ^{13}C NMR studies of polypeptides, proteins, and polysaccharides [see Saito (1986) for a review] have also provided information regarding their solid-state conformations and mobilities.

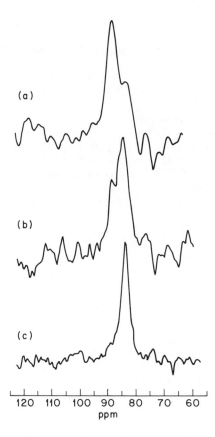

Figure 11.35 ▪ ^{15}N CPMAS NMR of nylon-6 homopolymer crystal forms: (a) mainly γ-nylon-6; (b) predominately α-nylon-6; (c) α-nylon-6. [Reprinted with permission from Powell et al. (1988).] δ^{15}N relative to crystalline glycine.

In addition, several 2D NMR techniques have recently been developed and applied to solid polymer systems [see Ernst et al. (1987) for details]. These techniques have been successfully implemented to study the compatibility of polymer blends (Caravatti et al., 1986; Mirau et al., 1989), the motions in polymer solids (Schaefer et al., 1983; Spiess, 1988; Maas et al. 1987; Kentgens et al., 1987), and the skeletal-carbon connectivity in insoluble polymer systems (Schaefer, 1988). It is hoped that application of 2D NMR techniques to solids will revolutionize the characterization of solid-state polymer structures, motions, and interactions, as has already occurred for the solution NMR of macromolecules (Bovey and Mirau, 1988).

References

Arimoto, H. (1964). *J. Polym. Sci. Part A* **2**, 2283.

Axelson, D. E. (1986). In *High Resolution NMR Spectroscopy of Synthetic Polymers in Bulk*, R. A. Komoroski, Ed., VCH, Deerfield Beach, Fla., Chapter 5.

Belfiore, L. A., Schilling, F. C., Tonelli, A. E., Lovinger, A. J., and Bovey, F. A. (1984). *Macromolecules* **17**, 2561.

Bloor, D. and Chance, R. R. (1985). *Polydiacetylenes*, NATO ASI Series E, Applied Science, Martin Nijhoff Publishers.

Bovey, F. A. and Jelinski, L. W. (1987). *Nuclear Magnetic Resonance*, Encyclopedia of Polymer Science and Engineering, Vol. 10, Second Ed., Wiley, New York, p. 254.

Bovey, F. A. and Mirau, P. A. (1988). *Accts. Chem. Res.* **21**, 37.

Bowmer, T. N. and Tonelli, A. E. (1985). *Polymer* (*British*) **26**, 1195.

Boye, C. A., Jr., and Overton, J. R. (1974). *Bull. Am. Phys. Soc.* **19**, 352.

Brereton, M. G., Davies, G. R., Jakeways, R., Smith, T., and Ward, I. M. (1978). *Polymer* (*British*) **19**, 17.

Bunn, A., Cudby, E. A., Harris, R. K., Packer, K. J., and Say, B. J. (1981). *Chem. Commun.* 15.

Caravatti, P., Neuenschwander, P., and Ernst, R. R. (1986). *Macromolecules* **19**, 1889.

Chance, R. R. (1986). *Encyclopedia of Polymer Science and Engineering*, Vol. 4, Second Ed., Wiley, New York, p. 767.

Danusso, F. and Gianotti, G. (1963). *Makromol. Chem.* **61**, 139.

De Rosa, C., Napolitano, R., and Pirozzi, B. (1986). *Polymer* (*British*) **26**, 2039.

Desborough, I. J. and Hall, I. M. (1977). *Polymer* (*British*) **18**, 825.

Dickerson, R. E. and Geis, I. (1969). *Structure and Action of Proteins*, Harper, New York.

Downey, M. J., Hamill, G. P., Rubner, M., Sandman, D. J., and Velazquez, C. S. (1988). *Makromol. Chem.* **189**, 1199.

Earl, W. L. and VanderHart, D. L. (1979). *Macromolecules* **12**, 762.

Earl, W. L. and VanderHart, D. L. (1982). *J. Magn. Reson.* **48**, 35.

Enkelmann, V. and Lando, J. B. (1978). *Acta Cryst.* **B34**, 2352.

Ernst, R. R., Bodenhausen, G., and Wokaun, A. (1987). *Principles of Nuclear Magnetic Resonance in One and Two Dimensions*, Oxford University Press, New York.

Evans, H. and Woodward, A. E. (1978). *Macromolecules* **11**, 685.

Farrar, T. C. and Becker, E. D. (1971). *Pulse and Fourier Transform NMR*, Academic Press, New York.

Fisher, E. W. (1957). *Z. Naturforsch.* **12a**, 753.

Fleming, W. W., Fyfe, C. A., Kendrick, R. D., Lyerla, J. R., Vanni, H., and Yannoni, C. S. (1980). In *Polymer Characterization by ESR and NMR*, A. E. Woodward and F. A. Bovey, Eds., ACS Symposium Series 142, Washington.

Flory, P. J. (1969). *Statistical Mechanics of Chain Molecules*, Wiley-Interscience, New York.

Garroway, A. N., Ritchey, W. M., and Moniz, W. B. (1982). *Macromolecules* **15**, 1051.

Geacintov, C., Schottand, R., and Miles, R. B. (1963). *J. Polym. Sci. Polym. Lett. Ed.* **1**, 587.

Gomez, M. A., Tanaka, H., and Tonelli, A. E. (1987a). *Polymer* (*British*) **28**, 2227.

Gomez, M. A., Cozine, M. H., Schilling, F. C., Tonelli, A. E., Bello, A., and Fatou, J. G. (1987b). *Macromolecules* **20**, 1761.

Gomez, M. A., Cozine, M. H., and Tonelli, A. E. (1988). *Macromolecules* **21**, 388.

Gomez, M. A., Tonelli, A. E., Lovinger, A. J., Schilling, F. C., Cozine, M. H., and Davis, D. D. (1989). *Macromolecules* **22**, in press.

Grenier-Loustalot, M.-F., and Bocelli, G. (1984). *Eur. Polym. J.* **20**, 957.

Groth, P. (1971). *Acta Chem. Scand.* **25**, 725.

Hall, I. H. (1980). *ACS Symp. Ser.* **141**, 335.

Hall, I. H. and Pass, M. G. (1977). *Polymer* (*British*) **17**, 807.

Harrah, L. A. and Zeigler, J. M. (1985). *J. Polym. Sci. Polym. Lett. Ed.* **23**, 209.

Harrah, L. A. and Zeigler, J. M. (1987). In *Photophysics of Polymers*, C. E. Hoyle and J. M. Torkelson, Eds.; *ACS Symp. Ser.* **358**, 482.

Holmes, D. R., Bunn, C. W., and Smith, D. J. (1955). *J. Polym. Sci.* **17**, 159.

Iwayanagi, S. and Mirua, I. (1965). *Rep. Prog. Polym. Phys.* **8**, 1965.

Jakeways, R., Ward, I. M., Wilding, M. A., Hall, I. H., Desborough, I. J., and Pass, M. G. (1975). *J. Polym. Sci. Polym. Phys. Ed.* **13**, 799.

Jakeways, R., Smith, T., Ward, I. M., and Wilding, M. A. (1976). *J. Polym. Sci. Polym. Lett. Ed.* **14**, 41.

Jelinski, L. W. (1986). In *High Resolution NMR Spectroscopy of Synthetic Polymers in Bulk*, R. A. Komoroski, Ed., VCH, Deerfield Beach, Fla., Chapter 10.

Jelinski, L. W., Dumais, J. J., Watnick, P. I., Bass, S. V., and Shepherd, L. (1982). *J. Polym. Sci. Polym. Chem. Ed.* **20**, 3285.

Keller, A. (1957). *Phil. Mag.* **2**, 1171.

Kentgens, A. P. M., de Boer, E., and Veeman, W. S. (1987). *J. Chem. Phys.* **87**, 6859.

Kuzmany, H., Rabolt, J. F., Farmer, B. L., and Miller, R. D. (1986). *J. Chem. Phys.* **85**, 7413.

Maas, W. E., Jr., Kentgens, A. P. M., and Veeman, W. S. (1987). *J. Chem. Phys.* **87**, 6854.

Mark, J. E. (1967). *J. Am. Chem. Soc.* **89**, 6829. [Also see: Abe, Y. and Flory, P. J. (1971). *Macromolecules* **4**, 219.]

Mathias, L. J., Powell, D. G., and Sikes, A. M. (1988). *Polymer (British)* **29**, 192.

Mencik, Z. (1975). *J. Polym. Sci. Polym. Phys. Ed.* **13**, 2173.

Miller, R. D., Farmer, B. L., Fleming, W., Sooriyakumaran, R., and Rabolt, J. F. (1987). *J. Am. Chem. Soc.* **109**, 2509.

Miller, R. L. and Holland, V. F. (1964). *J. Polym. Sci. Polym. Lett. Ed.* **2**, 519.

Mirau, P. A., Tanaka, H., and Bovey, F. A. (1989). *Macromolecules* **22**, in press.

Miyashita, T., Yokouchi, M., Chatani, Y., and Tadokoro, H. (1974). In Annual Meeting of Polymer Science Japan, Tokyo, preprint, p. 453. Quoted in Tadokoro, H. (1979). *Structure of Crystalline Polymers*, Wiley-Interscience, New York, p. 405.

Natta, G. and Corradini, P. (1960a). *Nuovo Cimento Suppl. 15*, **1**, 9.

Natta, G., Corradini, P., and Porri, L. (1956). *Atti Acad. Naz. Lincei Cl. Sci. Fis. Mat. Nat. Rend.* **20**, 718.

Natta, G. and Corradini, P. (1960b). *Nuovo Cimento Suppl. 15*, **1**, 40.

Natta, G., Corradini, P., and Bassi. I. W. (1960). *Nuovo Cimento Suppl. 15*, **1**, 52.

O'Gara, J. F., Jones, A. A., Hung, C.-C., and Inglefield, P. T. (1985). *Macromolecules* **18**, 1117.

Petraccone, V., Pirozzi, B., Frasci, A., and Corradini, P. (1976). *Eur. Polym. J.* **12**, 323.

Powell, D. G., Sikes, A. M., and Mathias, L. J. (1988). *Macromolecules* **21**, 1533.

Saito, H. (1986). *Magn. Reson. Chem.* **24**, 835.

Sandman, D. J., Tripathy, S. K., Elman, B. S., and Sandman, L. M. (1986). *Synthetic Metals* **15**, 229.

Schaefer, J. (1988). Presented at Rocky Mountain Spectroscopy Conference, Denver, Aug. 1988. [Also see Bork, V., and Schaefer, J. (1988). *J. Magn. Reson.*, **78**, 348.]

Schaefer, J. and Stejskal, E. O. (1976). *J. Am. Chem. Soc.* **98**, 1031.

Schaefer, J. and Stejskal, E. O. (1979). In *Topics in Carbon-13 NMR Spectroscopy*, Vol. 3, G. C. Levy, Ed., Wiley, New York, p. 283.

Schaefer, J., Stejskal, E. O., and Buchdal, R. (1975, 1977). *Macromolecules* **8**, 291; **10**, 384.

Schaefer, J., McKay, R. A., Stejskal, E. O., and Dixon, W. J. (1983). *J. Magn. Reson.* **52**, 123.

Schilling, F. C., Bovey, F. A., Tseng, S., and Woodward, A. E. (1983). *Macromolecules* **16**, 808.

Schilling, F. C., Bovey, F. A., Tonelli, A. E., Tseng, S., and Woodward, A. E. (1984). *Macromolecules* **17**, 728.

Schilling, F. C., Valenciano, M., and Tonelli, A. E. (1985). *Macromolecules* **18**, 356.

Schilling, F. C., Bovey, F. A., Lovinger, A. J., and Zeigler, J. M. (1986). *Macromolecules* **19**, 2660.

Schilling, F. C., Gomez, M. A., Tonelli, A. E., Bovey, F. A., and Woodward, A. E. (1987). *Macromolecules* **20**, 2954.

Schilling, F. C., Bovey, F. A., Lovinger, A. J., and Zeigler, J. M. (1989a). *ACS Adv. Chem. Ser.* in press.

Schilling, F. C., Lovinger, A. J., Zeigler, J. M., Davis, D. D., and Bovey, F. A. (1989b). Manuscript in preparation.

Shoji, A., Ozaki, T., Fujito, T., Deguchi, K., and Ando, I. (1987). *Macromolecules* **20**, 2441.

Spiess, H. W. (1985). *Adv. Polym. Sci.* **66**, 23.

Spiess, H. W. (1988). Presented at Am. Chem. Soc. Meeting, Toronto, Canada, June 6, 1988.

Stambaugh, B. D., Koenig, J. L., and Lando, J. B. (1979). *J. Polym. Sci. Polym. Lett. Ed.* **15**, 299; *J. Polym. Sci. Polym. Phys. Ed.* **17**, 1053.

Stellman, J. M., Woodward, A. E., and Stellman, S. D. (1973). *Macromolecules* **6**, 30.

Storks, K. H. (1938). *J. Am. Chem. Soc.* **60**, 1753.

Suehiro, J. and Takayanagi, M. (1970). *Macromol. Sci. Phys.* **B4**, 39.

Sun, D. C. and Magill, J. H. (1987). *Polymer (British)* **28**, 1245.

Tadokoro, H. (1979). *Structure of Crystalline Polymers*, Wiley-Interscience, New York.

Tanaka, H., Thakur, M., Gomez, M. A., and Tonelli, A. E. (1987). *Macromolecules* **20**, 3094.

Tanaka, H., Gomez, M. A., and Tonelli, A. E. (1988a). *Macromolecules* **21**, 2934.

Tanaka, H., Gomez, M. A., Tonelli, A. E., Chichester-Hicks, S. V., and Haddon, R. C. (1988b). *Macromolecules* **21**, 2301.

Tanaka, H., Gomez, M. A., Tonelli, A. E., and Thakur, M. (1989a). *Macromolecules* **22**, 1208.

Tanaka, H., Thakur, M., Gomez, M. A., and Tonelli, A. E. (1989b). *Macromolecules* **22**, in press.

Tanaka, H., Gomez, M. A., Tonelli, A. E., Lovinger, A. J., Davis, D. D., and Thakur, M. (1989c). *Macromolecules* **22**, in press.

Tanaka, H., Gomez, M. A., Tonelli, A. E., Chichester-Hicks, S. V., and Haddon, R. C. (1989d). *Macromolecules* **22**, 1031.

Till, P. H. (1957). *J. Polym. Sci.* **24**, 301.

Torchia, D. A. (1978). *J. Magn. Reson.* **30**, 613.

Turner-Jones, A. (1963). *J. Polym. Sci. Part B* **1**, 455.

Turner-Jones, A., Aizlewood, J. M., and Becket, D. R. (1964). *Makromol. Chem.* **75**, 134.

VanderHart, D. L. (1981). *J. Magn. Reson.* **44**, 117.

van Dongen, J. P. C. M., de Bie, M. J. A., and Steur, R. (1973). *Tetrahedron Lett.* **1371**.

Walsh, C. A., Schilling, F. C., Macgregor, R. B., Jr., Lovinger, A. J., Davis, D. D., Bovey, F. A., and Zeigler, J. M. (1989). *Macromolecules* **22**, in press.

Ward, I. M. and Wilding, M. A. (1977). *Polymer (British)* **18**, 327.

Weber, W. D. and Murphy, P. D. (1987). *Preprints PMSE Div. ACS* **57**, 341.

Wegner, G. (1980). *Discuss. Faraday Soc.* **68**, 494.

Yokouchi, M., Sakabibara, Y., Chatani, Y., Tadokoro, H., Tanaka, T., and Yoda, K. (1976). *Macromolecules* **9**, 26.

Zannetti, R., Manaresi, P., and Bazzori, G. C. (1961). *Chim. Indust. (Milan)* **43**, 735.

Index